Ronan O'Dowd © 2011
(200 pages)

Ronan O'Dowd's Designer Books in Photonic Engineering:

Photonics Handbook 1

"Photonics Handbook: Part 1 Broadband Fibres"

"Photonics Handbook: Part 1 Broadband Fibres"

Author

Ronan O'Dowd PhD SMIEEE is Professor Emeritus Photonic Engineering at UCD Dublin, Ireland where he taught and researched Optoelectronics and Photonics for three decades until 2010. He has several breakthrough papers in topics such as tunable semiconductor lasers and optical communications, including the millennium 2001 paper proving a dense comb of 2000 wavelength channels could be transmitted in a single fibre using the same semiconductor laser (ref *IEEE Jnl.S.T. Quantum Electronics Mar 2001*). Many of his students have proceeded to successful careers in academia and the photonics industry worldwide.

By the same author:

Physics Science of Action   **Gill and Macmillan 1984**

Tips to use this guidebook.

This series of books for photonic system designers will cover the subsystems that make up an optical communications link. These are the transmitter, fibre channel and receiver and since the fibre design sets the specifications and key criteria to be implemented at either end we tackle that first in this part 1 of the book.

The guide is formatted with *You Do* exercises and answers are provided. These are intended to be part of the learning process that will take the student to Engineering degree high level over what may be about a 30 hour degree module where 6 additional hours are set aside for practical work. The hardware link kit from OptoSci, where optical fibre dispersion etc can be measured over fibre reels, is very useful for laboratory explorations.

The later part 2 covering laser, detector and system design for broadband, can constitute a further full module.

"Photonics Handbook: Part 1 Broadband Fibres"

CONTENTS

Practical work may use the OptoSci kit which contains transmitter, fibre reel and receiver along with test equipment for dispersion, bit-error rate etc. Section 7 recommends sample experiments.

Notes:

You Do exercises should be attempted especially by the self-taught student using this guide and regardless of your confidence in your answer quality. Having then read the answer provided you should again attempt it.

Diagrams are simple line-style that the student should re-draw.

There are sample examination questions at the end using standard mathematical relations.

1   Design Targets

The greatest obstacle to broadband is dispersion. Data bits spread in time and overlap causing errors.

Radio communication uses frequencies of hundreds of kilohertz to megahertz in the form of electromagnetic waves that carry information by modulation. Modulation impresses information bits onto the waves.

Mobile and satellite systems use higher frequency microwaves at 1-10 GHz

At this x1000 times higher frequency the information carrying bandwidth or bw is in turn x1000 times greater, a fact of the science.

Infra-red light is at over 100 THz so we expect $10^{14}/10^{10}$ or bw capability over 10,000 wider again.

So what must we do to actually achieve this BROADBAND capability using light instead of microwaves?

Firstly we choose fibre optics over metal for many reasons. List of advantages of optical fibre over metal:

[1]   Low loss (0.2 dB / km); much greater span.
[2]   High bandwidth (many GHz.km); greater bit-rate.
[3]   Interference free unlike metal as glass is an insulator.
[4]   Low cross-talk as glass does not leak to nearby fibres.
[5]   Light weight (kg versus tonnes); easier installation in ducts and vehicles (e.g. aircraft / automobiles).
[6]   Small size (0.125 mm so many fibres per cable); suits existing ducts and cable pulling.

[7] Spark free; suits industrial environments.

[8] Glass raw material is silicate or silica, SiO_2, and is abundant unlike metals.

You Do...For you to do now

Ex 1: Draw a long glass rod and show light rays entering and then travelling down it by total internal reflection.

Show faster and slower light paths.

Is Snell's Law of refraction relevant anywhere in your figure?

Answer to Ex 1

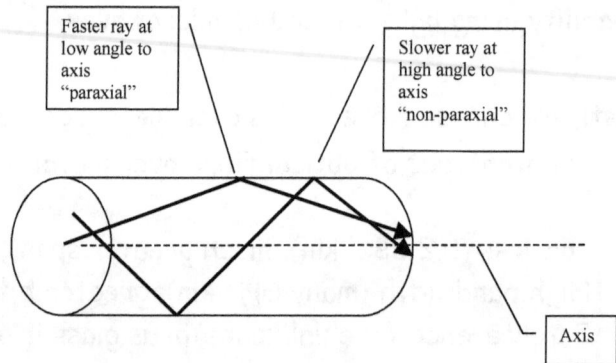

Faster ray at low angle to axis "paraxial"

Slower ray at high angle to axis "non-paraxial"

Axis

Figure 1A.

Snell's Law of refraction is relevant where the rays enter and exit the rod.

It also applies to the rays that reflect by total internal reflection TIR since these must strike the boundary at greater than the critical angle for refraction which is calculated for a ray that refracted out of the rod at 90 degrees.

You Do...For you to do next

Ex 2: Draw the slowest and fastest rays in the rod.

Use the glass refractive index of $n_1 = 1.5$ to estimate the path difference per metre length of rod where the diameter is 0.1 mm

Answer to Ex 2

The five figures below help visualise the ray that glances the boundary at just greater than critical angle C. Medium 1 is glass (index 1.5 or n_1) and medium 2 is air (index 1.0 or n_0).

The second figure shows how to calculate C using Snell's law

$$\sin 90 \deg / \sin C = n_1 / n_0$$

$$\rightarrow \quad C = \sin^{-1} n_0/n_1 = \sin^{-1}(1/1.5) = 42 \deg$$

The third figure traces that ray back to before it entered the rod and uses Snell again to calculate A the angle within which rays are accepted and later guided by TIR. Beyond that capture zone light may enter but is then lost from the guide by refraction. The glancing ray in the middle figure is at (90 – C) deg to the axis so that for entrance facet refraction

$$n_1 / n_0 = \sin A / \sin (90\text{-}C)$$

$$\sin A = 1.5 \sin (90\text{-}42) = 1.115$$

This result is greater than 1 and therefore called non-physical as it cannot happen. This is because the angle 48 deg exceeds the critical angle.

Rays at that divergence from the axis would never have entered in the first place. A lesson here is that the maths contains warning signs if we are alert to them.

The fourth figure shows that angle A actually defines in three dimensions or 3D the acceptance cone of light that can enter and be guided by TIR. The same cone applies by symmetry at the exit facet. Since later we will add a cladding glass around the core of the fibre whose index is only 1% lower than the main glass core the angle A is quite small in practice and hence transmitters and receivers must be extremely precisely aligned with optical fibres to avoid large coupling losses. That is a challenge for mechanical engineers. If the cladding had index $n_2 = 1.49$ and we repeat the above calculation we find $C = \sin^{-1} (1.49/1.5) = 83.4$ deg so that 90-C = 6.6 deg and that is well within the range of 42 deg for the air-glass entrance facet. In that case the third diagram shows that sin A/sin 6.6 = 1.5/1 giving $A = \sin^{-1} (1.5\sin 6.6) = 13.3$ deg and that result is now a fairly low aperture. Angle C within the guide is very large now (as a quick calculation by you using Snell can show) so all TIR rays are paraxial (meaning close to the axis) when a cladding glass is included in the design. We will therefore plan to create a glass cladding later as we proceed with improvements.

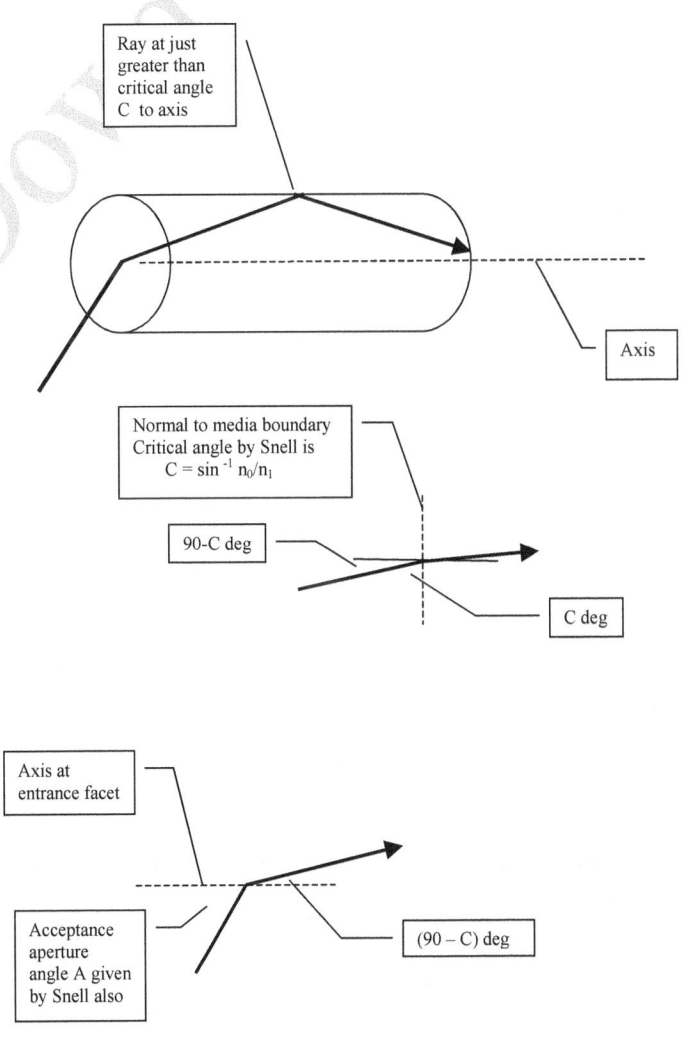

Ray at just greater than critical angle C to axis

Axis

Normal to media boundary
Critical angle by Snell is
$C = \sin^{-1} n_0/n_1$

90-C deg

C deg

Axis at entrance facet

Acceptance aperture angle A given by Snell also

$(90 - C)$ deg

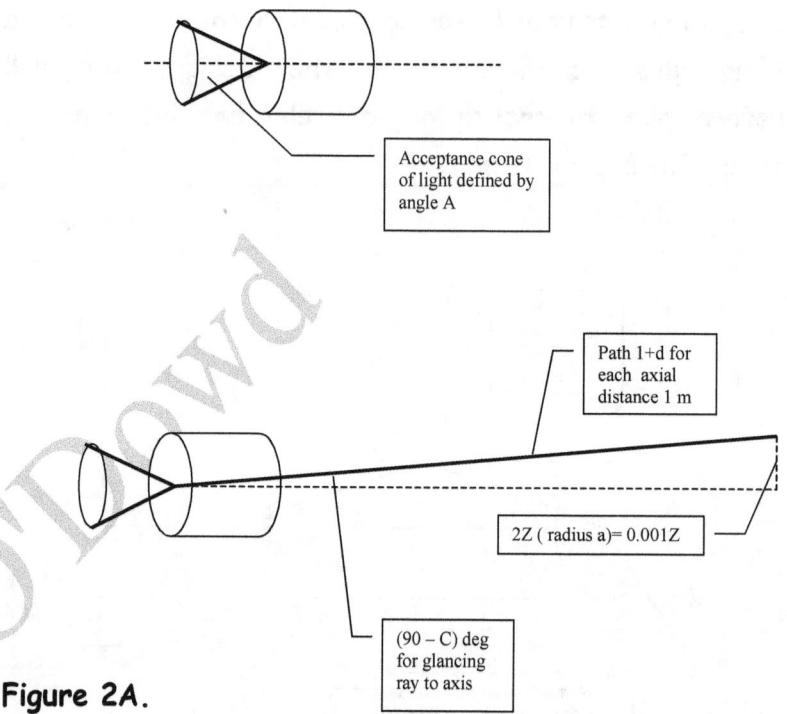

Acceptance cone
of light defined by
angle A

Path 1+d for
each axial
distance 1 m

2Z (radius a)= 0.001Z

(90 – C) deg
for glancing
ray to axis

Figure 2A.

The fifth figure shows a glancing ray path 1+d where d is the extra distance in metres travelled for each 1 m down the axis. Note the units there: metres per metre. The effective dispersion per metre of fibre, d, is calculated from

$$\cos(90-C) = 1/(1+d) \text{ or } \cos^{-1}[1/(1+d)] = 90-C$$

Using C for the core-cladding from above in this equation gives dispersion d, the path difference in metres per metre travelled down the axis.

Finally if we divide by speed of light c/n_1 the dispersion is in seconds for each metre down the axis.

For curiosity we may also find from this triangle the number Z of zig-zags or reflections for the ray per metre:

$$(1+d)^2 = 1^2 + (2aZ)^2$$

Note the 2aZ since for each reflection the ray then travels 2a across the guide to reach the boundary again. We can transpose each zig-zag segment of the path into a straight line for computation without altering the maths. This linear transformation can easily be shown using graph paper.

Low Attenuation Glass

A rod of conventional glass loses light so fast that after a few metres it is dark. Even a window pane which is that thick is impenetrable to light, visible or infra-red, IR.

The second important requirement after bandwidth is near lossless glass to achieve several km range. Ideally for inter-city communications we need > 100 km with minimum loss.

If the receiver (a well designed RX) can detect 100 times less light than is launched by the laser transmitter or TX then 99% less power is tolerable. The remaining 1% is 1/10 x 1/10 times so in decibels (ten times log base 10) that is minus 10 dB and then minus another 10 dB down or −20 dB total. This is because log 10 of 1/10 is -1; times 10 gives -10 dB.

You do...

Ex 3: Power on dB scale.

A semiconductor laser emits 2 mW of light power from each end of the chip. Translate to the commonly used photonic engineering decibel or dB units.

Answer to Ex 3

We take either 1 mW or 1 μW as reference power. For the former the laser emits 2 mW which is x2 times the reference and the $\log_{10}$ of that is 0.3 called 0.3 bel. There are 10 decibel to one bel so the power is 10x0.3 giving 3 decibels relative to a milliwatt. This is written 3 dBm where the m refers to a milliwatt, mW.

Alternatively relative to one microwatt or a 1 μW reference power the laser emits x2000 times and $\log_{10}$ of that is 3.3 giving 33 dBμ where the μ refers to a microwatt, μW.

Each additional 10 dB added means x10 times greater power in comparison to the reference so 30 dB added means 10x10x10 or x1000 times greater relative to the reference power.

1 dBm equals 30 dBµ and 3 dBm is the same as 33 dBµ. Remember that multiplication becomes addition in terms of logarithms. That is why we use the dB scale in the first place.

Since $\log_{10} 2 = 0.3$ or 3dB then doubling a power adds 3 dB.

The laser above emits at each chip end equally so for total power out of it add 3 dB giving 3+3=6 dBm or 33+3=36 dBµ in total.

Laser transmitters at **1.5 µm wavelength**, λ, are well developed using light emitting diodes made of compound semiconductors and not of silicon but rather semiconductor quaternary alloys of indium In, gallium Ga, arsenic As and phosphorous P.

Since frequency $\nu = c/\lambda$ we compute $3 \times 10^8 / 1.5 \times 10^{-6}$ giving a light frequency ν of 2×10^{14} Hz so this laser is at 200 THz or 200,000 GHz

So the capacity is 200,000 times broader than for a microwave system at 1 GHz. That is certainly broadband at the TX. A high speed response at the receiver RX is also needed so the photodiode and its amplifier circuit must be carefully designed too.

We also require range of several km.

The product of both these specifications, frequency and distance, in **GHz.km** (gigahertz kilometre) is therefore a measure of both important requirements in our target design.

Equally, the inverse **ns/km** (nanoseconds per kilometre) can be used but in that case the smaller the better.

You do...

Ex 4: Think about the units ns/km and figure out what they actually mean in the physical system.

Answer to Ex 4

As data travels down the fibre the existence of fast and slower pathways causes dispersion d calculated above. This is measured in nanoseconds or picoseconds spreading of the light pulses and the further it travels the greater the spread or dispersion. The dispersion after each kilometre is measured in nanoseconds per kilometre, ns / km. That is the inverse of the bandwidth times distance product in GHz.km.

Power Loss

Let us tackle the range problem first from the two targets, distance and bandwidth..

The attenuation or loss in glass is due to many factors that classify into absorption and scattering.

Attenuation is high at UV and shorter visible wavelengths due to Rayleigh scattering and slowly drops off according to 4th power of λ as we enter infra-red or IR until we reach 1.5 μm. After this it rises again by absorption as we enter the longer IR.

That is why the TX laser needs to be at *1.5 μm minimum loss.*

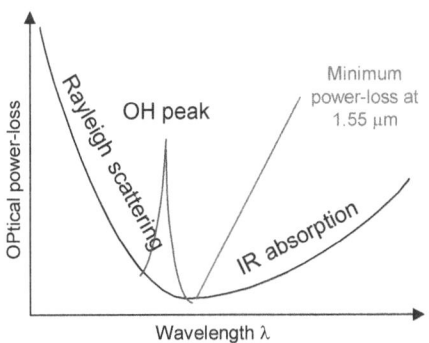

Figure 3A. Master power-loss curve for silicate glass SiO$_2$.

Rayleigh scattering, Figure 3A, falls with $1/\lambda^4$ as we move to longer wavelength from blue while after the red colours IR absorption takes over beyond 1.55 μm wavelength. There is a localised peak at 1.4 μm due to absorption by unavoidable, tiny, residual water content (OH-bond absorption); that peak separates the 1.3 μm or S-band (for short) from the 1.5 μm or C-band (for conventional). Compound semiconductor alloys InGaAsP of different elemental mix are needed to transmit at these different wavelengths and that science is called band-gap engineering. The optimum low-loss region around 1.55 μm is called the C-band for conventional and it can accommodate a comb of many closely spaced optical frequencies or wavelengths called dense wavelength division multiplexing DWDM. A single optical fibre may carry 50 DWDM channels or more in the C-band.

Chemical Vapour Deposition CVD

Fibre manufacture involves extremely pure gas ingredients unlike window panes that are made from treated sand. The gases carry elements Silicon and Oxygen that react in a flame to form pure SiO_2 or silicon dioxide which is glass. Other carrier gases bring dopants to the reaction like boron or phosphorous to alter the refractive index according to precisely programmed design. In that way we can create an INDEX PROFILE across the rod diameter of the so-called preform that results from the depositing solids (Figure 4A). The index profile formed by precise vapour deposition determines whether the fibre will be step-index multimode, graded-index multimode or monomode (single mode).

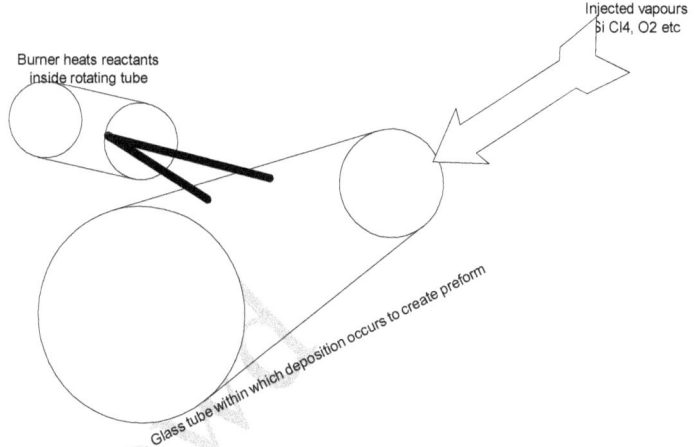

Injected vapours
Si Cl4, O2 etc

Burner heats reactants
inside rotating tube

Glass tube within which deposition occurs to create preform

Figure 4A. Glass perform grown by chemical vapour deposition CVD.

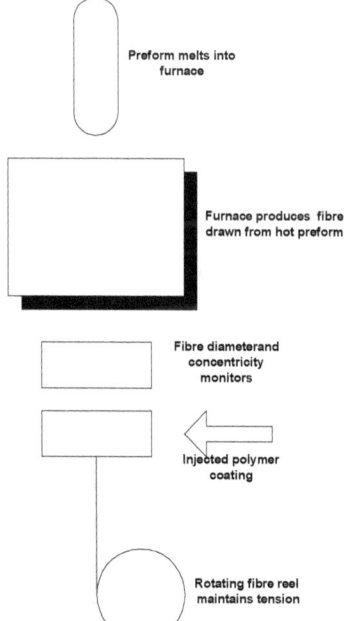

Preform melts into furnace

Furnace produces fibre drawn from hot preform

Fibre diameterand concentricity monitors

Injected polymer coating

Rotating fibre reel maintains tension

Figure 5A. Drawing tower to produce fibre reel of 1 or more kilometres from glass preform.

To draw the bulky preform into thin fibre a drawing tower is used (Figure5A). The cylindrical sample of about 1m long by several cm diameter grown by chemical vapour deposition CVD can be heated and drawn as a 0.1 mm strand onto a reel carrying several km length of fibre. This fibre has diameter 125 or 100 μm and its refractive index profile mimics that of the original bulk preform. Next the fibre is cabled for strength and protection from impurity ingress or mechanical impacts. Loose-tube cables have internal plastic grooves in an extruded central form that let the fibre sit without stress. Several fibres may be accommodated in each cable.

To join two reels of fibre a precision splicer is required as the 100 μm thick ends must match within <1 μm axial alignment accuracy (Figure6A).

When a re-usable joint is needed a precision mechanical connector that is demountable and of similar accurate capability is essential (Figure7A).

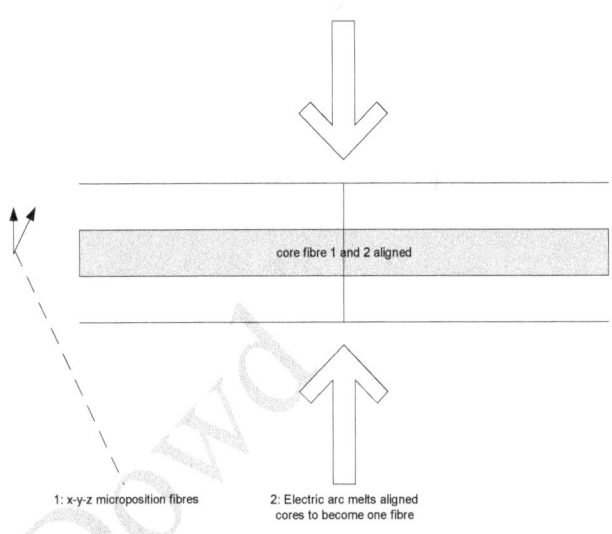

Figure6A. Fusion splicer for fibres; coupling loss 0.1 dB.

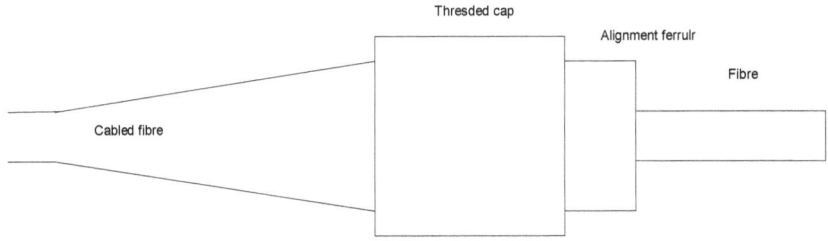

Figure 7A. Precision connector for de-mountable joints.

The reel or installed cable can be tested by optical time-domain reflectometry, OTDR, where a fast laser pulse of high power is launched and the back-scattered and reflected light is measured to identify cracks as well as the gradual fall-off in power with increasing range (Figure8A).

The underlying gradual fall-off is a logarithmic process measured in decibels per kilometre dB / km while reflection losses at couplers are in dB.

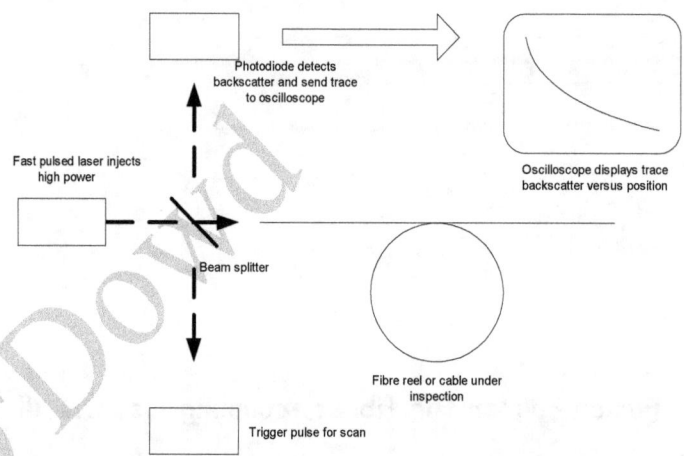

Photodiode detects backscatter and send trace to oscilloscope

Oscilloscope displays trace backscatter versus position

Fast pulsed laser injects high power

Beam splitter

Fibre reel or cable under inspection

Trigger pulse for scan

Figure 8A.   Optical time domain reflectometer OTDR.

You Do...

Show that a perform of 1 m long and 1 cm diameter can be drawn onto a reel of 100 μm diameter fibre with length 10 km.

You do...

Ex 5: Link design example with power-loss budget measured in dBm

A laser launches 4 dBm into a fibre with 0.4 dB/km loss at that wavelength. The span of 10 km reaches a detector with 3 dBμ sensitivity but with a 1 dB coupling loss. There are two connectors each with a 1.5 dB coupling loss. Allow for temperature and time degradation to assess final excess power if any.

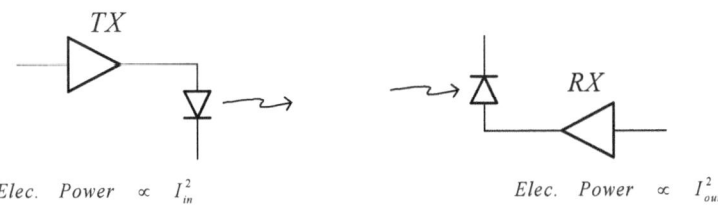

Figure 9A. Transmitter TX and receiver RX in fibre link.

At the fibre ends the TX has a laser diode emitting light while the RX has a photodiode receiving light. Electrical power depends on square of current but light power is proportional to current itself as electrons produce photons.

Answer to Ex 5

Launched at TX laser: +4 dBm

Reciever RX sensitivity: 3 dBμ or -27 dBm
Margin: 31 dB

Fibre loss for 10 km: 0.4x10= 4 dB
Connectors loss: 1.5x2= 3dB
Splice loss: no splices= 0 dB
Detector coupling loss: 1 dB
Allowance for temperature: 3 dB or factor 2 fall off
Allowance for time degradation   3 dB
Total attenuation: 14 dB
Excess power: 31-14 = 17 dB

Conclusion: each 3 dB means double the required power so there is vast power capacity here. A cheaper, less sensitive detector could be selected or, since sensitivity falls off with data rate, a much higher data speed could be used to improve broadband capacity as more customers join this part of the network or as they demand higher quality service.

○

Summary 1

The fibre we deploy must be of the highest calibre glass created by CVD in a fashion similar to silicon chips with the difference that we need silicon dioxide, called silicate or silica. While silicon or Si is a semiconductor the glass is an insulator, SiO_2. This extremely pure glass has exceptionally low attenuation, about 0.2 dB/km permitting several tens of km range without any repeater or booster. In that case a link of 100 km would have a loss of 0.2x100 = 20 dB and so the RX input power would be 100 times lower than that launched at the TX laser. A 3 mW laser could therefore deliver 30 μW at the detector in this example.

2   Rays and LP Modes

Rays give the direction in which the wavefront of linearly polarised light travels. This is called an *LP mode.*

Communications systems have three primary sub-systems. These are transmitter TX, fibre channel and receiver RX.
Each has its own response time and the inverse of that time is the sub-system bandwidth limit bw. The response times combine like Gaussian responses to give a lower full bw as follows:
System response time = [TX time ^{2} + channel time ^{2} + RX time ^{2}]$^{1/2}$
The overall response time as given here is called the root-mean-square, rms. The inverse is the system bandwidth bw.

So TWO BUDGET ANALYSES are needed in system design, power-loss and bandwidth.

You Do...

Ex 6: Link design example 2, bandwidth budget.

A link will use non-return to zero or NRZ data encoding where the system rise time is 0.7/data rate. It is a telephone cable at 140 Mbit/s with a 0.5 ns laser response. The total dispersion in the fibre span is 1.22 ns as measured for example with a laboratory OptoSci kit. The PIN detector has a 1 ns response. Perform a rise-time or bw budget.

Answer to Ex 6

Data rate:	140 Mbit/s
Required system rise time:	
for NRZ 0.7 / 0.14x10^9 =	5 ns
Laser rise time	0.5 ns
Fibre dispersion:	1.22 ns
Detector rise time:	1.0 ns
Sum of squares:	2.74
System root-mean-square rise time:	rms 1.66 ns

Conclusion: the response speed is fast and more than sufficient.

o

A POWER BUDGET is required to select the fibre for the system. This fibre must also satisfy the BANDWIDTH BUDGET so it is designed by engineers who grow the glass somewhat like silicon chips are made. The BROADBAND link will need fibre of excellent REFRACTIVE INDEX PROFILE so that the fastest rays of light and the slowest rays keep in pace and do not spread data time-wise. Alternatively we may let the fast rays only prevail with single mode operation.

What refractive index profile inside a tiny optical fibre will deliver broadband? The only sure way to answer that is to model the rays with mathematics. But a clue to what we expect the analysis to deliver (*and if it does not we will suspect our mathematics*) can be reasoned out as follows:

The dispersion d in ns/m or ns/km that we estimated for the primitive design example at Ex 2 occurred quite simply because the slow and fast rays or linearly polarised LP modes have different path lengths in the glass. Recall that the rod was 1 m x 0.1 mm. Let us call the diameter D = 0.1 mm and radius a will be half of that.

You Do...

Ex 7 :Reducing modal dispersion.

Re-draw the rod and rays again labelling fast and slow LP modes.

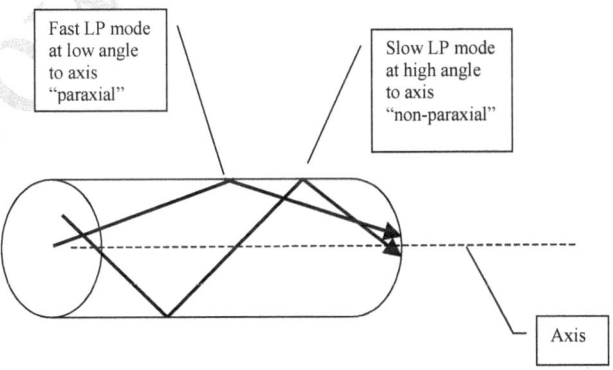

Figure 10A.   Fast and slow modes of the guide.

What property of glass could now speed up only the slow modes?

Answer to Ex 7

Speed of light in glass v = c/n where c is fixed at 3×10^8 so that refractive index n must be made lower to speed up slower modes. Where do these slow modes reside more often but fast paraxial modes do not?

The answer is near edges of the glass rod. Therefore we shall reduce n at and towards the edges of the guide, perhaps gradually like this:

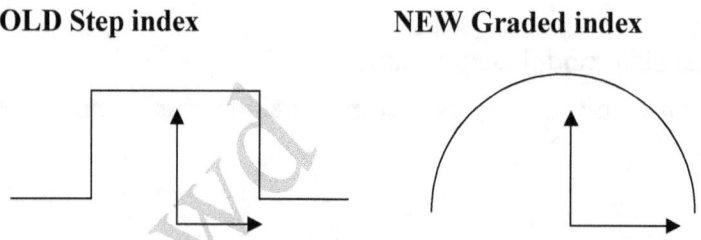

OLD Step index   **NEW** Graded index

Figure 11A. Refractive index profiles; horizontal arrows are radial position r from 0 to a, vertical arrows represent increasing index n(r) as a function of radial distance from the axis.

So a graded index GI instead of a step index SI profile should reduce the spread of mode speeds called the MODAL DISPERSION and the mathematical analysis should predict that kind of result.

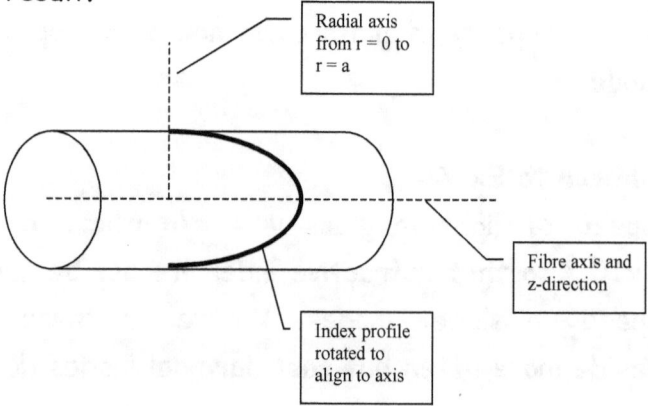

Radial axis from r = 0 to r = a

Fibre axis and z-direction

Index profile rotated to align to axis

Figure 12A. Graded index plot from last figure turned on its side and superimposed to line up with the fibre optic guide axis.

Another Improvement

Core-cladding guide: To protect the glass core itself from moisture and contamination a coating of polymer is laid during the drawing tower process (Figure 5A). But this protective polymer layer absorbs light so a second pure glass outer region, the cladding, is grown around the core during the CVD process. Then the polymer is coated on by the fibre drawing tower. This can avoid polymer absorption loss. We must therefore extend the profile by as much again, it turns out, and the index plot becomes:

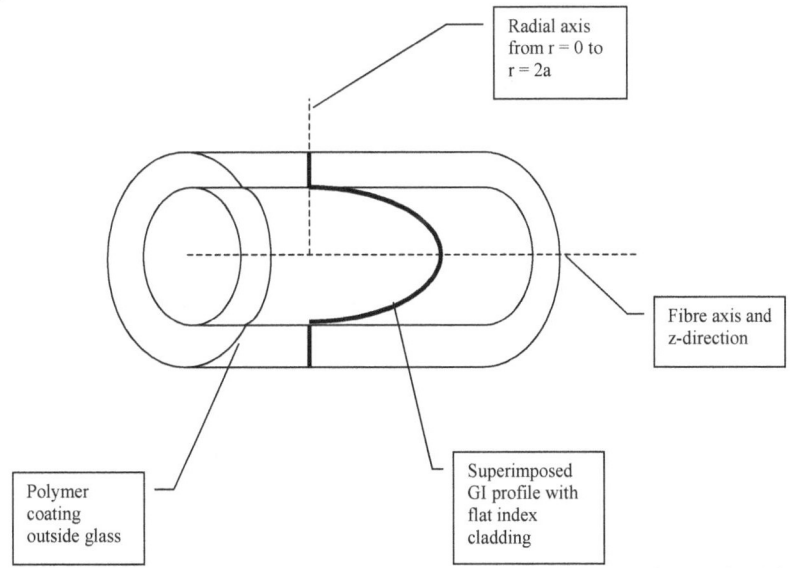

Radial axis from r = 0 to r = 2a

Fibre axis and z-direction

Polymer coating outside glass

Superimposed GI profile with flat index cladding

Figure 13A. Graded index fibre with flat index cladding and polymer coating.

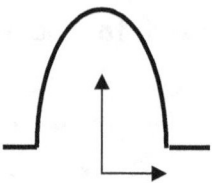

Figure 14A. The new profile has a GI core surrounded by a flat index cladding. Vertical arrow shows rising index and horizontal arrow is radial r from 0 to a.

You Do...

Ex 8: Re-read Ex 2. Calculate dispersion where n_1 or $n_{core} = 1.5$; n_2 or $n_{cladding} = 1.49$; n_0 or $n_{air} = 1$ and where the slowest ray enters the fibre from air at angle A deg to the axis.

Use Snell's law to find angle inside fibre for slowest LP mode and also the new critical angle for TIR.

What is the dispersion d?

Draw the cone of light defined by A that can enter the fibre from air using the above calculations by tracing a TIR ray backwards from inside to air outside the fibre end facet.

In practice the end facets are cleaved with a special tool to a shiny finish to minimise end coupling losses.

Answer to Ex 8.

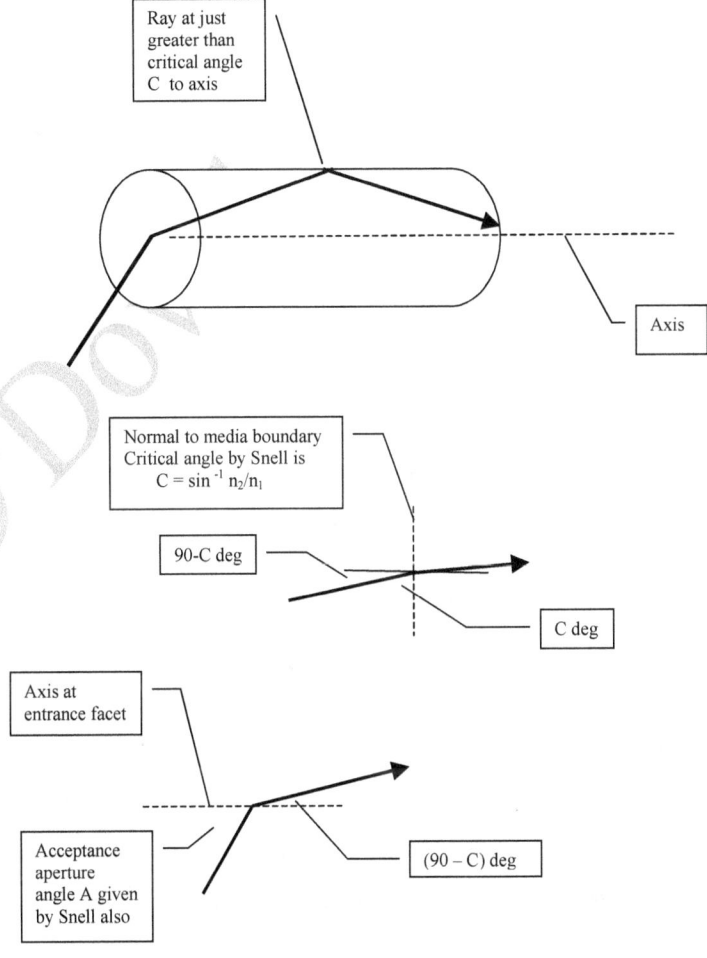

Ray at just greater than critical angle C to axis

Axis

Normal to media boundary
Critical angle by Snell is
$C = \sin^{-1} n_2/n_1$

90-C deg

C deg

Axis at entrance facet

Acceptance aperture angle A given by Snell also

$(90 - C)$ deg

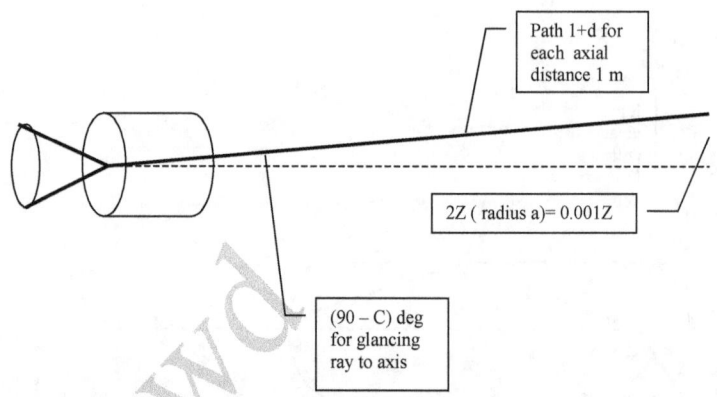

Path 1+d for each axial distance 1 m

2Z (radius a)= 0.001Z

(90 – C) deg for glancing ray to axis

Figure 15A Calculating "primitive" dispersion by ray method.

Re-read Ex 2 first.

Recalculate: $C = \sin^{-1}(n_2/n_1 = \sin^{-1}(1.49/1.5) = 83.4$ deg
$$90 - C = 6.6 \text{ deg}$$

By Snell at the entrance facet
$$A = 13.3 \text{ deg}$$
$$\cos(90-C) = 1/(1+d) \text{ or } \cos^{-1}[1/(1+d)] = 90-C$$
[Or simply $\sin C = 1/(1+d) = n_2/n_1$ etc]

Now cos 6.6 = 0.9933 so 1+d = 1/0.9933 = 1.0067 metre per metre. This gives d=0.0067 metre per metre
Alternatively $d = 0.0067/(c/n_1) = 34$ ps/m or 34 ns/km
Bandwidth distance product is the inverse: 0.03 GHz.km

Finally $(1+d)^2 = 1^2 + (2aZ)^2$ and $2a = 50\times10$ -6 therefore:
$$Z = (0.52\times10^6)^{1/2} = 748$$
This result is large because the diameter is so small.

○

Summary 2

The fibre should be chemically grown SiO_2 glass to achieve high purity. The refractive index profile should have a higher index at the axis gradually falling towards the edge and with a lower, flat index cladding around the core. This design reduces modal dispersion considerably improving bw.

A cladding of glass at a flat index about 1 % below the axial value contributes and separates the lit-up core from the polymer coating that protects the extremely pure glass from light absorbing contaminants to sustain loss budget over many years. The cladding has a great effect on the critical angle. In turn the acceptance angle is reduced but as a result many higher modes are not guided. This in conjunction with the reduced index provided by a graded profile means that further from the axis the dispersion is greatly reduced. That is GI multimode fibre and is an improvement on SI multimode profile in terms of bandwidth by a factor of $x \, 10^{5}$

Ray and Wave Optics

Ray optics was used to reach this point but we need to move to WAVE OPTICS now to get a design engineer's perspective. A ray gives the direction in which an electromagnetic WAVEFRONT moves. A plane wavefront therefore has parallel rays. A spherical wavefront has radial rays emanating from a point. The WAVE EQUATION from Maxwell's theory of electromagnetism describes the motion through space (x,y.z) in time (t).

That equation uses a vector operator in space so let us briefly review a few fundamentals. From electromagnetic theory we know that electric field and voltage are related:

$$E_r = - \delta V/dr$$

When r is a vector in xyz space and i,j,k are unit vectors in these three directions this becomes:

$$E = iE_x + jE_y + kE_z$$
$$= - (i\delta/\delta x + j\delta/\delta y + k\delta/\delta z)V$$

The operator in brackets is called grad for gradient of the voltage or also del and symbolised as Δ and it is a vector since i, j and k are unit vectors in x, y and z directions:

$$\Delta V = \mathbf{grad}V = (i\delta/\delta x + j\delta/\delta y + k\delta/\delta z)V$$

An inverted delta can be used if you prefer.

The dot product of Δ with itself is Δ^2:

$$\Delta^2 = \Delta.\Delta = \delta^2/\delta x^2 + \delta^2/\delta y^2 + \delta^2/\delta z2$$

This is shown by simply multiplying out the bracketed terms $(i\delta/\delta x + j\delta/\delta y + k\delta/\delta z)$ with itself to produce six terms and noting that three terms are zero being cross products of orthogonal unit vectors like ixj etc and that $i^2 = j^2 = k^2 = 1$.

The common form of Maxwells equation for an E field in the vertical or y-direction is:

$$\Delta^2 \mathbf{E_y} = \mu_0\varepsilon_0 \, \delta^2\mathbf{E_y}/\delta^2 t^2$$
$$= (\delta^2/\delta\mathbf{x}^2 + \delta^2/\delta\mathbf{y}^2 + \delta^2/\delta\mathbf{z}^2) \, \mathbf{E_y}$$

The Δ^2 operator may be written in other co-ordinate systems such as spherical but the one we will opt for is cylindrical to match the shape of optical fibres.

In cylindrical polar co-ordinates r, ϕ, z we find:

$$\Delta^2 = \delta^2/\delta\mathbf{r}^2 + 1/\mathbf{r} \, . \, \delta/\delta\mathbf{r} + 1/\mathbf{r}^2 \, . \, \delta^2/\delta\phi^2 + \delta^2/\delta\mathbf{z}^2$$

Another expression we will deploy is the speed of light in a medium like glass. The speed of light c in vacuum or air is related to the progress of electrical and magnetic effects. In a dielectric it is the electrical permittivity ε and magnetic permeability μ of the medium that determine the speed v:

$$1/c^2 = \mu_0\varepsilon_0$$

In glass where r means "relative to vacuum" this becomes:

$$1/\mathbf{v}^2 = \mu_0\mu_r\varepsilon_0\varepsilon_r$$

For transparent materials $\mu_r = 1$ (approx.) so dividing these two equations we get:

$$\mathbf{v}/\mathbf{c} = (1/\varepsilon_r)^{1/2}$$

This ratio of speeds in the medium c/v is also its refractive index. Therefore relative permittivity ε_r is related to refractive index n:

$$n = (\varepsilon_r)^{1/2}$$

A further expression that we need to account for power loss when light enters the glass from air was established by Fresnel. The reflected and transmitted portions are related to refractive index by:

$$R = [(n-1)/(n+1)]^2 \quad \text{and } T = 1-R$$

You Do...
Ex 9: Draw plane and spherical wavefronts and the associated directional rays.
Ansswer to Ex 9

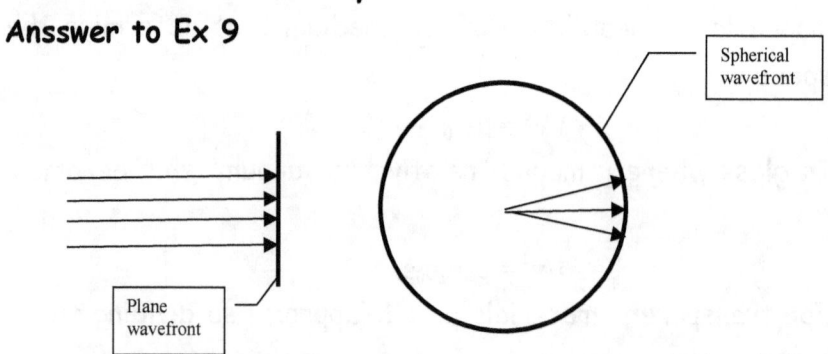

Figure 16A. Wavefronts and rays.

3   Mathematical Model

WKB (Wentzel Kramers Brillouin) Method

This technique was applied in 1973 to optical waveguides by Gloge and Marcatili at Bell Labs. We require the dispersion or bandwidth or impulse response for the cylindrical optical waveguide. Start with the wave equation where we assume index variation is negligible over a wavelength distance.

You Do...

Ex 10: Show index variation is negligible over a wavelength distance to be the case using the following data: radius fibre 25 μm, core index 1.51, cladding index 1.5 and laser wavelength 1550 nm.

Ans Ex 10 at end handbook.

$$\nabla^2\psi = n^2\varepsilon_0\mu_0\frac{\delta^2\psi}{\delta t^2} \tag{1}$$

This wave equation 1 relates the variation in space, del-squared, of the optical field psi or ψ (left hand side, lhs) to its variation in time (right hand side, rhs). The light will be contained inside the core of a cylindrical fibre so that the operator del-squared on the lhs will be more convenient in cylindrical polar co-ordinates giving equation 2.

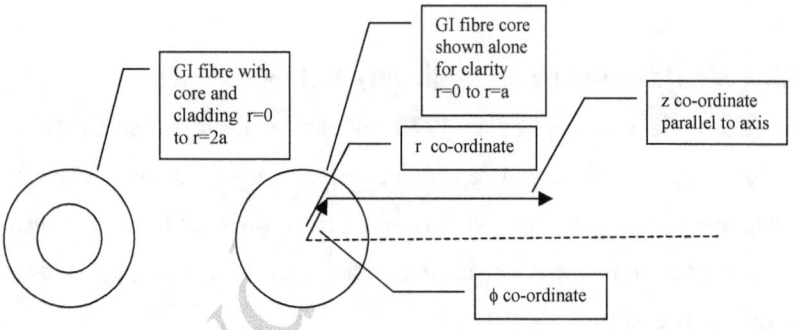

Figure 1. The fibre has core and cladding, left, but for brevity and clarity we will show the core alone as we proceed. Cylindrical polar co-ordinates in the fibre are r,φ,z. Move out radially r inside the core of radius a, then swing around polar or azimuthal by φ, and travel down parallel to the axis distance z.

$$\Rightarrow \frac{\delta^2 \psi}{\delta r^2} + \frac{1}{r}\frac{\delta \psi}{\delta r} + \frac{1}{r^2}\frac{\delta^2 \psi}{\delta \phi^2} + \frac{\delta^2 \psi}{\delta z^2}$$
$$= n^2 \varepsilon_0 \mu_0 \frac{\delta^2 \psi}{\delta t^2} \tag{2}$$

Now we have the variation in space on the lhs in terms of radial r, polar or azimuthal φ and finally axial position z or distance down the fibre from the launch point. The rays that indicate direction of travel bounce back and forth by TIR at the core interface with the cladding. Outward and returned waves interfere as is the nature of light. Since there is cylindrical symmetry, as the light progresses we should expect the patterns of bright and dark created by the waves as they interfere to have cylindrical symmetry themselves.

Postulate solution: $\psi = F(r)\cos(\nu\phi)e^{i(\omega t - \beta z)}$ **MODE**

This has the expected cylindrical symmetry in the radial r and azimuthal or ϕ senses where the integer ν describes the polar or azimuthal mode pattern or its periodicity. There are ν pattern repeats as we swing round the circle from 0 to 2π radians so this is the **azimuthal mode number**.

The cosine in respect of this polar position means the brightness rises and falls periodically in a circle while the pattern also varies radially. Such patterns also appear for example on the skin of a vibrating drum. We insert this form of field, the mode, into the wave equation next.

$$\Rightarrow \frac{d^2F}{dr^2} + \frac{1}{r}\frac{dF}{dr} + \left(n^2k^2 - \beta^2 - \frac{\nu^2}{r^2}\right)F = 0 \tag{3}$$

You Do...

Ex 11: Differentiate the postulated solution, MODE, once and then twice with respect to each of the co-ordinates r, ϕ and z. Insert into the wave equation 2 and tidy up to get (3).

Ans Ex11 at end handbook.

Hint:

$$\psi = F(r)\cos(\nu\,\phi)e^{i(\omega t-\beta z)}$$

$$\frac{\delta\psi}{\delta r} =$$

$$\frac{\delta^2\psi}{\delta r^2} =$$

$$\frac{\delta^2\psi}{\delta\phi^2} =$$

$$\frac{\delta^2\psi}{\delta z^2} =$$

$$\Rightarrow \frac{d^2F}{dr^2}+\frac{1}{r}\frac{dF}{dr}+\left(n^2k^2-\beta^2-\frac{\nu^2}{r^2}\right)F = 0$$

Both cylindrical and spherical symmetry are common in engineering and science. In the WKB method it was found that solving this equation (3) in linear x-space is tedious but a transformation to exponentially or log compressed space greatly eases the mathematics. Next transform to a log type of space:

$$r = ae^x \quad \text{and hence} \quad dr = ae^x dx$$

$$\Rightarrow \frac{d^2F}{dx^2}+\left(\kappa^2 a^2 e^{2x}-\nu^2\right)F = 0 \tag{4}$$

Where $\qquad \kappa^2 = n^2k^2-\beta^2$

You do...

Ex 12: Use partial differentiation to get from Eq (3) to Eq(4)

Ans Ex12 at end handbook.

Observe the form of this equation (4); it suggests some kind of oscillatory solution. This is because it resembles the second order differential equation for simple harmonic motion:

$$\frac{d^2F}{dx^2} + (constsnt)F = 0$$

Therefore we give the solution amplitude and phase properties.

Use $\qquad F(x) = A(x)e^{is(x)}$ for amplitude A and phase S of the field.

Take second derivative of this product (with respect to) wrt x and insert into (4) to get (5):

$$\Rightarrow \quad \frac{d^2A}{dx^2} + 2i\frac{dA}{dx}\frac{dS}{dx} - A\left(\frac{dS}{dx}\right)^2 + iA\frac{d^2S}{dx^2} + \left(x^2a^2e^{2x} - v^2\right)A = 0 \qquad (5)$$

Break into real and imaginary parts and tidy up to get:

$$\Rightarrow \quad \begin{cases} \dfrac{dS}{dx} = \left(\kappa^2a^2e^{2x} - v^2\right)^{\frac{1}{2}} & (6) \\[3mm] \dfrac{d}{dx}\left[A^2\dfrac{dS}{dx}\right] = 0 \quad \Rightarrow \quad A = C\left[\dfrac{dS}{dx}\right]^{-\frac{1}{2}} & (7) \end{cases}$$

$$\Rightarrow \quad A(x) = \frac{C}{\left(\kappa^2a^2e^{2x} - v^2\right)^{\frac{1}{4}}} \qquad (8)$$

$$(6) \Rightarrow \quad S(x) = \int_{x_1}^{x}\left(\kappa^2a^2e^{2x} - v^2\right)^{\frac{1}{2}}dx \qquad (9)$$

Return to "linear" r-space from log type of space using:

$$r = ae^x \quad \text{and hence} \quad dr = ae^x dx$$

This gives: $\qquad A(r) = \dfrac{C}{\left[\left(n^2k^2 - \beta^2\right)r^2 - v^2\right]^{\frac{1}{4}}} \qquad (10)$

and
$$S(r) = \int_{1}^{r} \left[\left(n^2 k^2 - \beta^2 \right) - \frac{v^2}{r^2} \right]^{\frac{1}{2}} dr \qquad (11)$$

The phase S in $F(x) = A(x)e^{is(x)}$ should be real to produce an oscillatory solution, otherwise we get an exponentially decaying result called an evanescent wave as a result of $i^2 = -1$. For a sustained field or guided wave the expression in equation (11) must therefore be real; this passes to imaginary when the square root term in parenthesis passes through zero and goes negative. The transition from guided mode occurs at:

"Turning points"......
$$\left[n^2(r)k^2 - \beta^2 \right] - \frac{v^2}{r^2} = 0 \qquad (12)$$

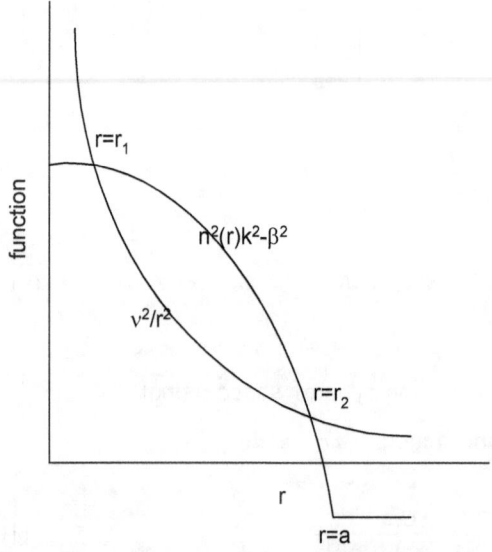

Figure 2. Plots to identify turning points.

To examine this we can draw plots for $\left[n^2(r)k^2 - \beta^2\right]$ and $\dfrac{v^2}{r^2}$ in order to identify where the first is above the second plot (Figure 2) so that the difference is positive. Reading inside the square brackets the first plot resembles the index profile squared times a constant and then pull it down by β-squared. The second plot is a constant over r-squared. This Figure 2 produces two values r_1 and r_2 defining an annular ring within which there is a real phase and hence a guided mode or ray, where the latter is the direction of travel for that mode (Figure 3).

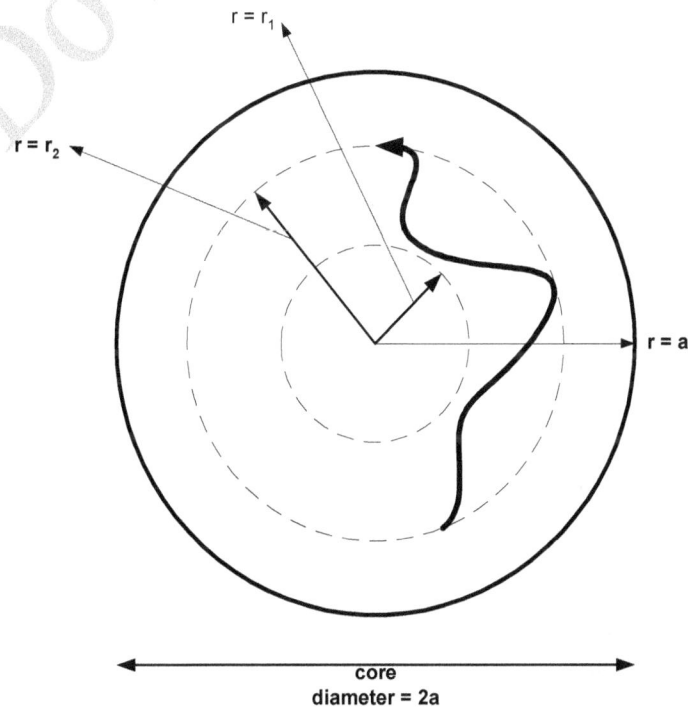

Figure 3. Azimuthal path of a guided ray within core section. The ray is spiralling down the fibre z-axis always contained inside the turning points for that mode r_1 and r_2.

In Figure 3 we see how each different mode will have different turning points as given by solution to equation (12). This means there is a dark spot at the centre for some modes, whenever r_1 is non-zero. There is also a dark ring outside r_2 if it is less than a. This is actually an interference pattern created by all the reflected waves that propagate down the core by diffraction.

Using equations (10) and (11) in the postulated solution for equation (2) namely Mode $\psi = F(r)\cos(v\,\phi)e^{i(\omega t - \beta z)}$ along with $F(r) = A(r)e^{is(r)}$ gives the result for the guided field:

$$\psi = \frac{1}{2}A(r)\{\exp[i(\omega t - \beta z - v\phi + S(r))] + \exp[i(\omega t - \beta z + v\phi + S(r))]\} \qquad (13)$$

You Do...
Ex 13: Produce equation (13) from (2) using (11).
Ans Ex 13 at end handbook.

○

The full standing wave should include by addition the negative square root from equation (11) and hence:

$$\psi = A(r)\cos(v\phi)\cos[S(r)]e^{i(\omega t - \beta z)} \qquad (14)$$

You do...

Ex 14: Produce Eq(14) from Eq(Mode) and (13)

Ans Ex 14 at end handbook.

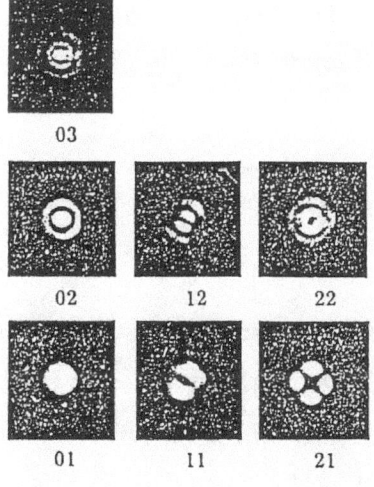

Figure 4   Linearly polarised LP modes $LP_{\nu\mu}$ (As measured by Stolen in 1976 after optically isolating each mode). The fundamental mode LP_{01} is at bottom left.   Study carefully and relate each to the mode integers.

The permitted patterns in Figure 4 are indicated each by two integers that are subscripts for that LP mode. When a ray traverses a full radial path from r_1 to r_2 and back the phase should return through an integer number μ of 2π, this being the second or **radial mode number**; additionally there are two 90 degree reflections at the turning points totalling another π. Hence we may write the phase condition:

$$2\int_{r_1}^{r_2}\left[n^2(r)k^2 - \beta^2 - \frac{v^2}{r^2}\right]^{\frac{1}{2}} dr = (2\mu + 1)\pi \qquad (15)$$

$\Rightarrow$ WKB *eigenvalue equation* including the second radial integer for the mode μ becomes:

$$\Rightarrow \qquad \int_{r_1}^{r_2}\left[n^2(r)k^2 - \beta^2 - \frac{v^2}{r^2}\right]^{\frac{1}{2}} dr = \left(\mu + \frac{1}{2}\right)\pi \qquad (16)$$

The two integers v and μ that describe a selected mode pattern are inserted into Eq (16) and the solution provides the propagation constant β. This process can be repeated for other index profiles $n(r)$ or for other wavelengths contained in k.

SUMMARY 3

LP modes are the solutions to the wave equation in the cylindrical guide and these allowed light patterns have a set of maxima and minima or bright and dark spots. Each mode pattern $LP_{\nu\mu}$ has cylindrical symmetry like the acoustic patterns on the skin of a cylindrical drum. Each LP mode has two integers that describe its pattern, the azimuthal ν and the radial μ giving maxima totals simply ν and $2\mu+1$ and in polar and radial directions respectively.

4   Dispersion Mechanisms

The WKB eigenvalue equation is so called because we insert fibre profile, laser wavelength, selected mode numbers and solve for propagation constant β of that mode. Then the distribution of β-values gives the modal dispersion i.e. arrival time in nanoseconds of fastest versus slowest modes after each km length of guide. Knowing the dispersion allows us to calculate the bandwidth.

Recall: Dispersion (ns/km) = Inverse bandwidth x distance product (1/GHz.km)

Modal Dispersion

At Bell Laboratories in the last century much research was done to produce the design for the first viable optical fibres satisfying commercial loss and bandwidth requirements. We still use the Gloge-Marcatili approach and their expression for a general index profile fibre is:

$$\left.\begin{array}{ll} n(r) = \left[1 - 2\Delta\left(\dfrac{r}{a}\right)^{\gamma}\right]^{\frac{1}{2}} & \cdots\cdots \quad r < a \\ \quad\quad = n_1\left[1 - 2\Delta\right]^{\frac{1}{2}} & \cdots\cdots \quad r \geq a \end{array}\right\} \qquad (17)$$

Here $\Delta = (n_1^2 - n_2^2)/2n_1^2$ which measures the relative drop in refractive index from core to cladding interface or the difference squared across the core profile.

This Δ approximates to $(n_1-n_2)/n_1$ and is therefore below 1%.

The index is assumed to fall from n_1 at $r = 0$ to n_2 at $r = a$ and γ (power exponent) is **profile shape-factor** or rate of fall-off. This was found by the researchers to be optimised at

$$\gamma_{opt} \approx 2(1 - 1.2\Delta) \tag{18}$$

Beyond the core the second part of equation (17) holds and shows the profile is flat at the interface value, also given by the first equation when $r = a$. (Check this yourself by putting $r = a$ in the first equation).

Since Δ is very small (try typical values for n_1 and n_2) Eq(18) yields a shape factor $\gamma = 2$ for optimum graded index profile multimode fibre. When we insert $\gamma = 2$ into Eq(17) the result is the typical expression for a parabola so a **parabolic index profile** is the optimum shape for multimode fibre.

For a single λ source like a good quality laser this was then shown to give the dispersion for graded index (G.I) fibre

$$\left[\Delta\tau_{opt} \approx \frac{Ln_1}{8c}\Delta^2 \right] \qquad \text{G.I. modal dispersion} \tag{19}$$

What this Eq(19) represents is the difference in time on the lhs between travel by the fastest and slowest modes over a length L of fibre.

You Do...

Ex 15: Insert typical values as used previously into Eq(19) to get dispersion in parabolic index guide over each km of fibre.

Ans Ex 15 at end handbook.

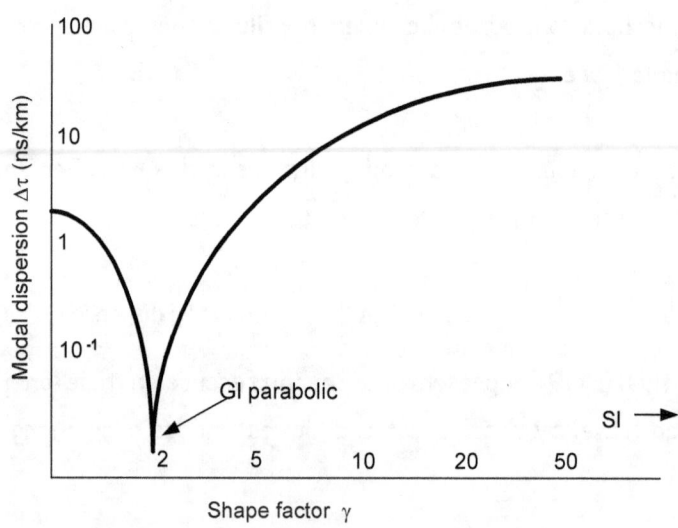

Figure 5   Modal   dispersion plot versus index profile shape factor. Observe the sharp cusp at GI parabolic represents improvement of 10^4 over SI.

This fibre dispersion due to modes may be re-calculated for other profiles by using a range of shape factors and the results are plotted in Figure 5. The γ-value is the ordinate and ranges from very low through $\gamma = 2$, GI parabolic, all the way to $\gamma = $ infinity for step-index SI as provided by Eq(17). This last point is because for any $r < a$ the fraction r/a to the power of infinity is zero so the *index is flat* at n_1 inside the core and then steps down abruptly to n_2 at the interface where $r = a$.

You Do...Ex: Show this easily using Eq(17) for core and cladding.

The plot is spread over a very wide range of resultant dispersions, so much in fact that a vertical log axis is used. There is a very sharp cusp at $\gamma = 2$ where the dispersion in ns/km is minimum. At the far right where γ approaches infinity, or step-index fibre, the dispersion reaches an ultimate value. Between the optimised parabolic GI profile and the SI case the dispersion improves by a whopping 100,000 so that bandwidth for GI parabolic is 10^5 times greater when we learn to grow this optimised profile. The challenge there following the maths is computer control of the precision valves that gradually open/close during glass perform growth.

Total Dispersion

There are further contributions to the spreading of the light pulses in the guide.

Total fibre dispersion comprises:

Modal dispersion as above

Profile dispersion where the profile is optimised, say parabolic, at one wavelength only but looks a different shape, not fully optimised, at others because index is wavelength dependent.

Wave-guide dispersion where dimensions of the guide play a role somewhat as happens when microwaves travel down a metal conduit.

Material dispersion, that for silica glass is *zero at 1300 nm* as was discovered by measuring the refractive index of glass over many colours. It was found that chromatic dispersion reverses (blue faster than red instead of red faster than blue, see Figure 6) at a specific wavelength depending on the material. This chromatic effect can overwhelm modal dispersion, particularly in GI fibre where the modal effect is greatly reduced or in monomode fibre where it is eliminated.

Chromatic dispersion is the combination of material along with the smaller profile effect above since both are colour dependent.

In classical optics of the nineteenth century anomalous dispersion was investigated and the plots of n versus λ recorded. The plot can be differentiated to yield data for material dispersion $\Delta\tau$:

$$\left[\Delta\tau = \frac{L\lambda}{c}\left(\frac{d^2n}{d\lambda^2}\right)\Delta\lambda \right]$$ Material dispersion   (20)

The difference in time of travel for longer versus shorter wavelengths was found to depend on the second rate of change of index with wavelength as shown in Eq(20). It also increased with the spread of wavelengths in the light source $\Delta\lambda$ and obviously with the physical length of the fibre link L or more specifically with L/c which is the time to travel L at speed c. Furthermore, as we move to longer wavelength λ the dispersion degrades. Eq (20) summarises all of this.

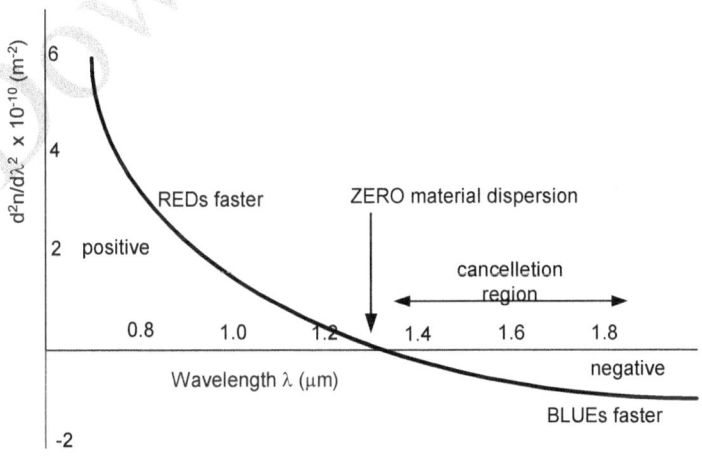

Figure 6   Material dispersion of SiO₂ glass. Zero occurs at 1.3 μm. RED is used for longer and BLUE for shorter wavelengths but all is in IR region.

Having so many effects turns out to be an advantage because single mode fibre eliminates the modal effect and the remaining chromatic (negative) and waveguide dispersion (positive) can be offset against each other at **1550 nm** (lowest loss in Figure 3A) to create **dispersion-shifted fibre** (see advanced fibres and Figure 13 later).

TABLE 1

Source	SI	GI	Monomode
LED λ=850 nm $\Delta\lambda$=50 nm	10/500 3	1000/500 3	100,000/500 2.5
Laser λ=850 nm $\Delta\lambda$=2 nm	10/25,000 3	1000/25,000 3	100,000/25,000 2.5
Laser l=1.3 μm $\Delta\lambda$=2 nm	10/100,000 1.0	1000/100,000 1.0	100,000/100,000 0.7
Laser l=1.55 mm DFB type	Library Exercise	Exercise	Exercise

Table 1   Dispersion and attenuation for various systems. The key to the data is: Modal dispersion/Material dispersion (both in MHz km) followed by Attenuation (dB km-1). The 1.55 μm laser TX case is left as an ambitious research exercise and can be done for monomode and dispersion shifted fibre.

Mode Cut-off Mechanisms

The guided LP modes may be investigated further with the help of Figure 2 that we used to find the turning points. We visualise the two graphs moving relative to each other and to the axis as shown below. The $n^2(r)k^2-\beta^2$ graph will move up as β reduces and down as β increases. This produces two extremes where the mode is said to be cut-off. An allowed range of β values therefore exists depending on profile n(r) and wavelength (contained in k).

The first case depicted in Figure 7a is where we consider modes with increasing v. At a limiting value the v^2/r^2 plot rises above the other and no turning point can be found. The annular ring has shrunk entirely. Such modes are not guided.

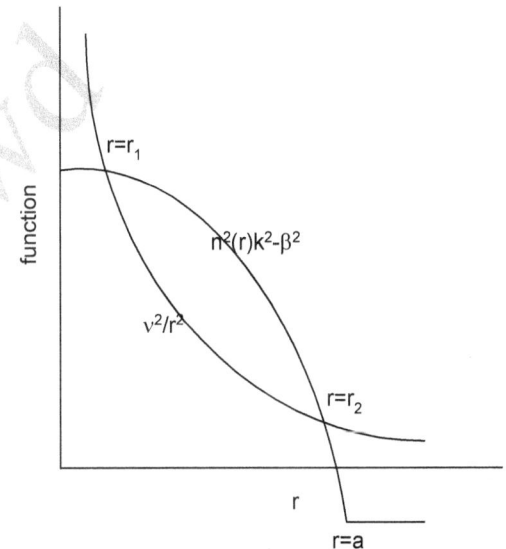

Figure 2 (Repeat) Finding the turning points r_1 and r_2 by the WKB method.

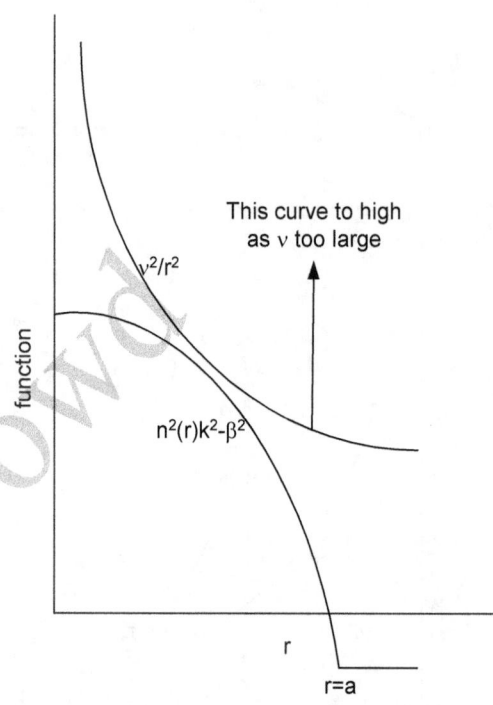

Figure 7a Cut-off mechanism 1. The v^2/r^2 curve has moved up with increasing mode number v until no intersection occurs. Modes with that value v or greater cease to be guided.

In the next two cases the β value is allowed to rise to a maximum β_{max} and then fall to a minimum β_{min}. In Figure 7b the whole plot resembling profile squared has gone below axis so no turning points are possible. In Figure 7c that curve has risen entirely above axis and again such modes are cut-off. In this case another phenomenon appears; the third intersection r_3 means a new region to the right permits guided propagation and tunneling of light from the inner guidance region takes place. Such "evanescent modes" radiate energy to the cladding since r_3 exceeds a and after a few tens of metres and are said to be lossey.

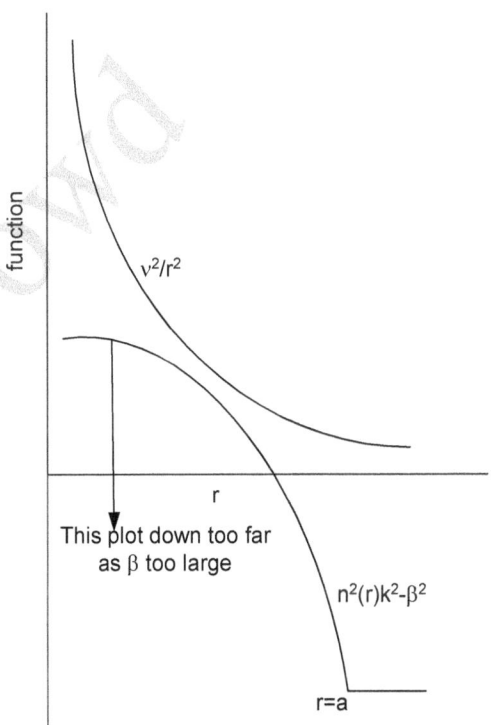

Figure 7b   Cut-off mechanism 2. The plot resembling the index profile squared moves down with increasing β until no intersection occurs. This situation must occur when the whole curve is below the axis and describes maximum β.

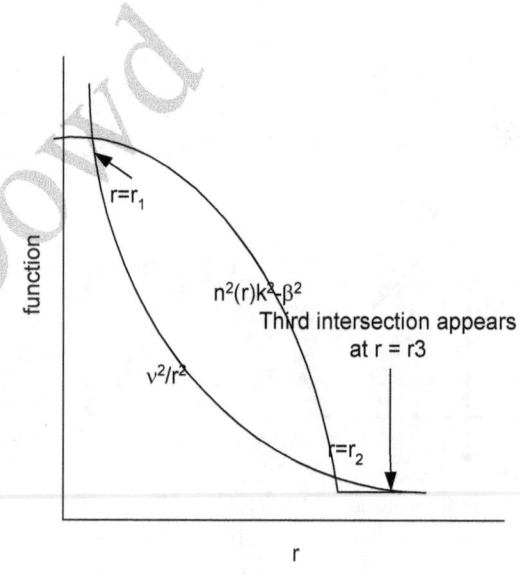

Figure 7c   Cut-off mechanism 3. A third intersection r_3 occurs when the whole curve rises above the axis as when b decreases too much. This case describes minimum β.

You Do...

Ex 16: By studying the figures above for cut-off and imagining the graphs moving up or down so that all the curve is above or all is below the axis...

Show $\qquad \beta_{min} = n_2 k \qquad$ **cut-off condition**

Show $\qquad \beta_{max} = n_1 k \qquad$ **cut-off condition**

Hence $\qquad n_2 k \quad < \quad \beta \quad < \quad n_1 k \qquad\qquad$ (21)

Ans Ex 16 at end handbook.

○

Number of Guided Modes

The LP modes of the fibre carry power and since they are eigen-solutions to the wave equation they all have equal energy. To see how much total energy we must address the question of how many modes exist in a multimode fibre?

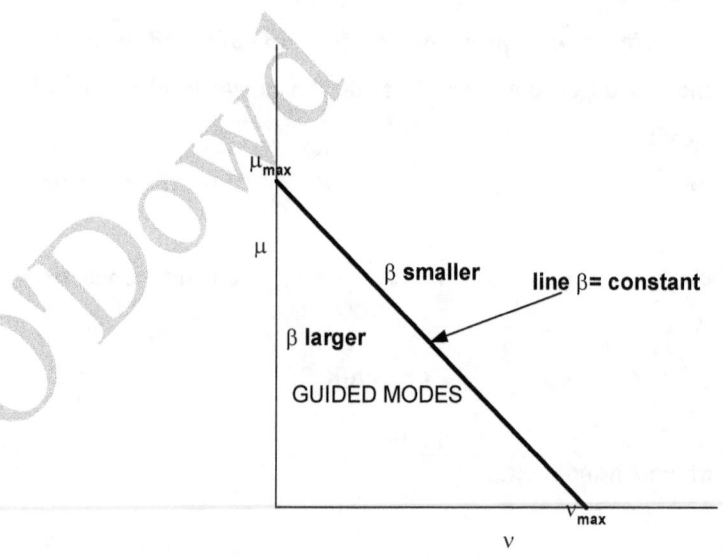

Figure 8 Guided LP modes depicted in mode number space.

Since each LP mode is defined by two integers we can draw a 2D chart, Figure 8, which depicts each mode as a dot with its ν number on the horizontal axis and μ number on the vertical. There turns out to be a line above which the mode does not exist as the integers rise above an allowed limit for propagation constant β. Recall we have already found the permissible range of β.

When we now count the dots under that line the result is the number of modes in the guide.

That count is essentially the same as the area under the curve which we get by integration. Hence the number M of modes, with β in the permitted range, equals the integral of vertical μ with respect to horizontal v:

$$M(\beta) \approx 4\int_0^{v_{max}} \mu\, dv$$

$$= \frac{4}{\pi}\int_0^{v_{max}} \int_{r_1}^{r_2} \frac{1}{r}\left\{\left[n^2(r)k^2 - \beta^2\right]r^2 - v^2\right\}^{\frac{1}{2}} dr\, dv \qquad (22)$$

But for completeness we have included a *multiplier 4* since there are both sine and cosine possibilities that are equally valid or periodic at the starting point Eq (2) and furthermore there can be two polarisation states of the same mode. So the integral in Eq (22) is immediately expanded by substituting for μ using the WKB eigenvalue Eq (16).

Using $n(R)k = \beta$ and keeping in mind the range for β given by Eq (21), namely $n_2 k < \beta < n_1 k$, we get:

$$M(\beta) = \int_0^R \left[n^2(r)k^2 - \beta^2\right]r\, dr \qquad (23)$$

Recall
$$n(r) = \left[1 - 2\Delta\left(\frac{r}{a}\right)^\gamma\right]^{\frac{1}{2}} \qquad \ldots\ldots \quad r < a$$
$$= n_1[1 - 2\Delta]^{\frac{1}{2}} \qquad \ldots\ldots \quad r \geq a \qquad (24)$$

with
$$\Delta = (n_1^2 - n_2^2)/2n_1^2 \qquad (25)$$

Here we must square $n(r)$, so by Eq (24) the exponent becomes 2γ, and then integrated (see Ex 16 below and end of book) so that

$$\Rightarrow \qquad M(\beta) = \frac{\gamma}{\gamma + 2}(n_1 ka)^2 \Delta\left[\frac{n_1^2 k^2 - \beta^2}{2n_1^2 k^2 \Delta}\right]^{\left(\frac{\gamma+2}{\gamma}\right)} \qquad (26)$$

61

$$\Rightarrow \qquad \beta = n_1 k \left[1 - 2\Delta\left(\frac{M(\beta)}{N}\right) \right] \qquad (27)$$

$$N = \left[\frac{\gamma}{(\gamma + 2)}\right](n_1 ka)^2 \Delta \qquad (28)$$

Here we have renamed the general $M(\beta)$ simply as N, the number of modes below the line for the limiting value of β and therefore the total number.

Let us for convenience define the fibre **V-parameter**:

$$\left(V \equiv n_1 ka\sqrt{2\Delta}\right)$$

This along with eq. (28) $\qquad \Rightarrow \qquad N = \frac{\gamma}{2(\gamma + 2)}V^2 \qquad (29)$

The last equation now provides tidy expressions for number of modes for various profiles:

$$\left.\begin{array}{l} \gamma \rightarrow \infty \quad (step-index) \quad \cdots \quad N = \frac{1}{2}V^2 \\[2mm] \gamma = 2 \quad (parabolic) \quad \cdots \quad N = \frac{1}{4}V^2 \end{array}\right\} \qquad (30)$$

Observe that our optimised GI fibre with parabolic profile has only half as many propagating modes. Since the modes are eigen-solutions to the wave equation they carry equal energy. Hence only half as much power is carried by the GI fibre. That is the price we must pay for 100,000 times higher bandwidth.

You Do...

Ex 17: Square Eq (24), insert result into Eq (23) and use the known max and min values for β to produce, after integration, Eq (25).

Hint: reverse the order of integration and use the tabulated intgrand of arc-sine from standard integral tables.

Answer Ex 17 at end handbook.

LP or Linearly Polarised Modes in SI Fibre

A deeper examination reveals that each LP mode is in fact the superposition of two waves, one with strong magnetic (HE) and the other strong electric character (EH) and that the mode numbers for these are related to those of our selected LP mode as shown below:

$$HE_{v+1,\mu} \quad + \quad EH_{v-1,\mu} \quad \Rightarrow \quad LP_{v\mu}$$

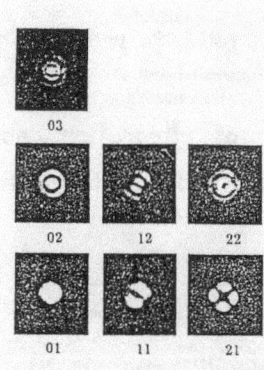

Figure 4 (Repeat). The LP modes as measured by Stolen.

Recall Eq.(2) again, for LP modes:

$$\psi = F(r)\cos(v\phi)e^{i(\omega t - \beta z)} \tag{31}$$

As before

$$\frac{d^2F}{dr^2} + \frac{1}{r}\frac{dF}{dr} + \left(K^2 - \frac{v^2}{r^2}\right)F = 0 \tag{32}$$

Where now we have defined

$$K^2 = n^2(r)k^2 - \beta^2 \tag{33}$$

The form of Eq.(32) is well known in mathematics and its solutions are cylinder functions. This does not surprise us since our light guide is cylindrical. The known cylinder functions are called Bessel and

Hankel. Of these two the Bessel function $J_v(Kr)$ remains finite at the co-ordinate origin $r = 0$ and is therefore the appropriate cylinder function solution to eq.(32) for the <u>core</u> region:

$$F(r) = AJ_v(\kappa.r) \quad \cdots \quad |r| < a \tag{34}$$

For the special case of <u>step index or SI fibre</u> we will use for the core region $\kappa = K$ with $n(r) = n_1$ in eq.(33) so that we now get

$$\kappa = \left(n_1^2 k^2 - \beta^2\right)^{\frac{1}{2}} \tag{35}$$

For the <u>cladding</u> (postulate infinite extent) the solution is different. Here $n_2 < n_1$ and $n_2 k < \beta < n_1 k$ for guided modes. Introduce $\gamma = K/i$ (n)t shape factor!!) so

$$\gamma = \left(\beta^2 - n_2^2 k^2\right)^{\frac{1}{2}} \tag{36}$$

and the cylinder function must have imaginary argument since the bracket term must be negative.

Furthermore for guided modes $F(r)$ vanishes as $r \to \infty$. These two characteristic properties indicate the modified Hankel function $K_v(\gamma r)$ to describe the cladding field:

$$F(r) = BK_v(\gamma r) \quad \cdots \quad |r| > a \tag{37}$$

Note
$$\kappa^2 + \gamma^2 = \left(n_1^2 - n_2^2\right)k^2 \tag{38}$$

Compare
$$V = ka\left(n_1^2 - n_2^2\right)^{\frac{1}{2}} \tag{39}$$

This latter is the so-called V-parameter for the optical fibre.

$$\Rightarrow \quad \left[(\kappa a)^2 + (\gamma a)^2 = V^2\right] \tag{40}$$

At the core-cladding discontinuity the field and its derivative are continuous:

$$\left.\begin{array}{rcl} AJ_v(\kappa.a) & - & BK_v(\gamma a) & = & 0 \\ \kappa AJ_v'(\kappa.a) & - & \gamma BK_v'(\gamma a) & = & 0 \end{array}\right\} \tag{41}$$

Solution by determinant method:

$$\kappa J_v'(\kappa.a)K_v(\gamma a) = \gamma J_v(\kappa.a)K_v'(\gamma a) \tag{42}$$

We may use from the mathematical handbook known relations for cylinder functions Eq (43):

$$\left.\begin{array}{rcl} J_v'(x) & = & \dfrac{v}{x}J_v(x) & - & J_{v+1}(x) \\[2mm] and \quad K_v'(x) & = & \dfrac{v}{x}K_v(x) & - & K_{v+1}(x) \end{array}\right\} \tag{43}$$

The previous equation (42) now becomes

$$\left[\; \kappa J_{v+1}(\kappa.a)K_v(\gamma a) = \gamma J_v(\kappa.a)K_{r+1}(\gamma a) \; \right] \tag{44}$$

This equation (44) is the eigenvalue equation for the LP modes.
It is used simultaneously with equation (40), where (γa) and $(\kappa.a)$ are variables while V is the fibre parameter and v is the mode number, to produce the propagation parameter β for that mode. The last step also requires equations (35) and (36).

You Do...

Ex 18: Show that Eq(40) and Eq(44) solved simultaneously with the help of Eq(35) and Eq(36) can deliver the propagation constsant β for a selected LP mode.

Ans Ex 18 at end handbook.

o

Single-mode Fibres

Cylinder function computer sub-routines permit solution to the LP mode eigenvalue equation (44) by numerical methods. The result is the propagation constant β for each permitted mode and the spread of these represents the modal dispersion in a SI fibre. As the guide's core-diameter is reduced the number of permitted modes falls according to Eq(30) $N = V^2/2$ and we would expect that for $V^2 = 2$ or $V = 1.414$ the result should be a lone mode or $N = 1$.

In fact the step-index or S.I. fibre becomes single mode (derived below) when

$$V < 2.405 \tag{45}$$

The reason for this paradox is that the starting assumption to derive Eq(30) had a slight simplification that does not matter when there are hundreds of modes as happens with typical multimode operation. We simplified $2\mu+1$ to 2μ. In the single mode case that assumption is invalid.

Mode Cut-off and the V-parameter

The cut-off condition whereby the argument in equation (36) turns from imaginary to real and the energy gets radiated instead of guided is

$$\gamma = 0 \tag{46}$$

Hence equation (40)

$$V^2 = (\kappa.a)^2 + (\gamma a)^2$$

gives

$$V_c = \kappa.a \tag{47}$$

Taking this as argument for the Bessel function in the eigenvalue eq.(44) and assuming $\gamma a \ll 1$ in the modified Hankel function approximations I and II (also from the mathematical handbook)...

$$
\left.
\begin{array}{lll}
\text{I} & K_v(x) = & \ln\!\left(\dfrac{2}{x}\right) & \cdots\ v = 0 \\[2ex]
and\ \text{II} & K_v(x) = & \dfrac{(v-1)!}{2}\!\left(\dfrac{2}{x}\right)^2 & \cdots\ v \geq 1
\end{array}
\right\}
\tag{48}
$$

We obtain using I:

$$\text{I} \qquad J_1(V_c) = J_0(V_c)/\left[V_c \ln\!\left(\dfrac{2}{\gamma a}\right)\right]$$

(49) Hence

$$J_1(V_c) = 0 \quad \text{as } \ln(\text{infinity}) = 0. \tag{50}$$

This is the cut-off condition for modes with $v = 0$

When $v \geq 1$ we obtain using II:

$$\text{II} \qquad \dfrac{2v}{V_c} J_v(V_c) - J_{v+1}(V_c) = 0 \tag{51}$$

Use from the mathematical handbook the functional relation:

$$V_c J_{v-1}(V_c) + V_c J_{v+1}(V_c) = 2v J_v(V_c) \qquad (52)$$

$$\Rightarrow \quad J_{v-1}(V_c) = 0 \quad \text{cut-off condition for modes with } v \geq 1 \qquad (53)$$

For the LP_{11} mode in particular this indicates $\quad J_0(V_c) = 0$

$$\underline{V_c = 2.405} \qquad (54)$$

This is telling us that the LP_{11} mode becomes cut-off as the V-parameter is reduced below this critical value 2.405 and therefore only the LP_{01} mode remains giving us single mode operation or monomode fibre.

We defined the fibre V-parameter $\quad (V \equiv n_1 k a \sqrt{2\Delta}) \quad$ so that by reducing the fibre diameter 2a sufficiently we must arrive at single mode propagation. Inserting V = 2.405 and typical n_1, n_2, and λ values in that definition will show that the fibre is monomode below about 10 μm diameter.

You Do...

Ex 19: From V parameter definition and knowing the associated single mode condition estimate the radius of the fibre where λ = 1.5 μm, n_1 = 1.5, n_2 = 1.49

Ans Ex 19 at end handbook.

○

The single mode fibre radius answer to Ex 19 is 3.3 μm or 6.6 μm diameter. Compare multimode fibre with 50 μm core diameter. By varying Δ the monomode answer can be up to 10 μm.

Cut-off V-parameter

The above discussion was for SI fibre but the whole analysis from Eq(32) to here would need to be repeated for other index profiles each with their own shape factor γ. In fact computational methods would be required for this due to the complexity when the core index is no longer fixed at a constant n_1. Nonetheless this has been performed and the result is plotted in Figure 9.

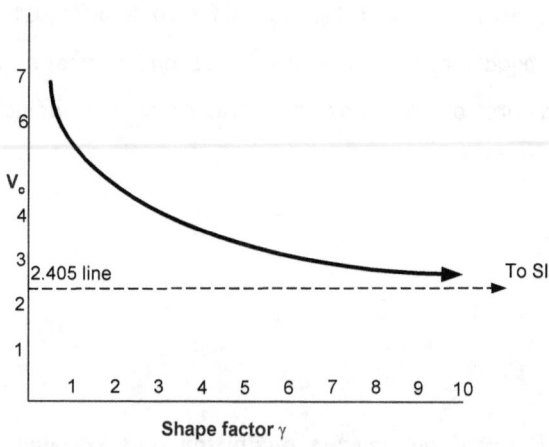

Figure 9. Cut-off V-parameter versus profile shape-factor; SI fibre to right.

Cut-off Wavelength of the Fibre

The wavelength λ is itself contained within the definition of V-parameter so that we should expect that at a certain wavelength the V-parameter ensures that the mode becomes cut-off. We find that after designing the fibre to be monomode and then by using shorter and shorter wavelength lasers ultimately a second mode appears. That is called the *cut-off wavelength* because at longer than that λ the second mode gets eliminated. This λ_c can be measured by launching variable wavelength light from a monochromator into a reel of fibre and observing the output light pattern with a IR camera. At longer than cut-off wavelength there is only the LP_{01} mode but at shorter than λ_c the LP_{11} mode appears and the centre of the optical near-field dims as depicted in Figure 10.

Figure 10(below). As the injected light wavelength shortens below cut-off wavelength a second mode appears so the pattern is a superposition of LP_{01} and LP_{11}.

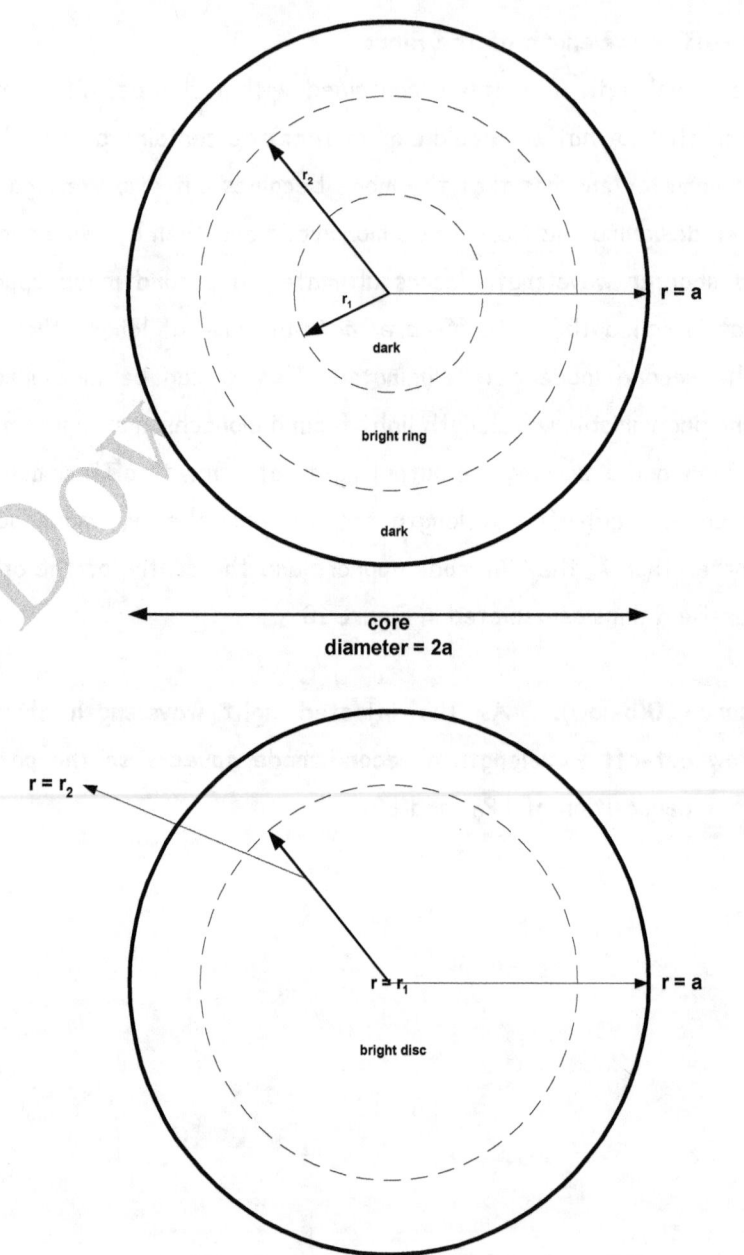

5   Advanced Fibre Designs

We have learned to design the ultimate single mode fibre at a selected wavelength but when a range of many wavelengths are used simultaneously as separate channels in the one fibre the system is called WDM for wavelength division multiplexing. In such cases only one of the comb of channels would be optimised while the others would suffer. Therefore we need to return to the analysis of trade-offs to see how the full set of wavelengths might be made to propagate with an equal but minimal dispersion rather than just one being optimised while all others degrade. The outcome is a more advanced index profile such as depicted for comparison in Figures 11 and 12. These are the result of very detailed mathematical scrutiny and are deployed in dense WDM systems. The whole C-band in the 1550 nm region of the optical spectrum is filled with channels 50 GHz apart in optical frequency. This gives the so called ITU (International Telecommunications Union) comb of channels.

Figure 11. Non-dispersion-shifted profile and triangular profile *zero* dispersion-shifted fibre optimised at low-loss wavelength 1550 nm rather than prior minimum-dispersion λ at 1300 nm

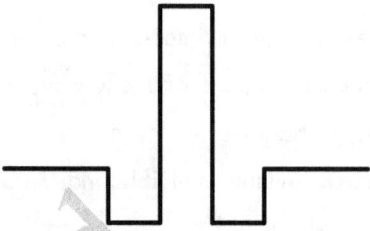

Figure 12   W-profile for dispersion-flattened fibre accommodating a comb of C-band channels with a compromise dispersion value that is flat across all chosen wavelengths.

Polarisation Maintaining Fibre

There a many situations where it is advantageous to deploy fibre that transmits one polarisation only. For example sensors often utilise polarised light. Coherent communications (like optical FM) systems are another case in point. By creating more stress across the diameter in one preferred direction it is found that a single polarisation is suppressed with the other remaining. That stress is achieved in a number of ways depicted in Figures 13 and 14.

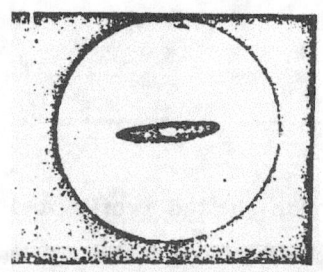

Figure 13. Polarisation maintaining fibres; elliptical core.

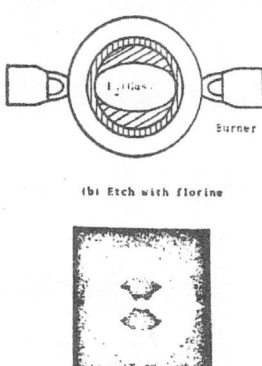

(b) Etch with florine

(d) Final cross-section

Figure 14. Polarisation maintaining fibre with bow-tie core. The perform growth process is shown revealing the etching stage with burner and fluorine gas to create the stress-creating shaped core. (University of Southampton).

In one case the glass preform is grown with an elliptical core and in the other with a bow-tie shape. This causes the asymmetrical stress required.

We can represent the optical power flowing down the z-axis as comprising x and y polarisations orthogonal to one another and to the axis. The rate of transfer of power from P_x to P_y take place with distance z at a rate we will designate h so that the equations (55) summarise what is happening.

$$\frac{dP_x}{dz} = h(-P_x + P_y) \quad \text{and} \quad \frac{dP_y}{dz} = h(P_x - P_y) \tag{55}$$

$$\cdots \quad \eta = \tanh(hz) \tag{56}$$

The form of these coupled equations is familiar in mathematics where the solution takes the form of equation (56) and the hyperbolic tangent trend becomes evident.

Alternative Glasses

The conventional optical glass is SiO2, namely silica or silicate but other glasses have been explored to see if longer links could be created without expensive repeaters. One such candidate was fluoride glass and its performance is shown in Figure 17 for comparison. While promising in the laboratory the fluoride glass is however unstable in air and not used for that reasons in broadband cables.

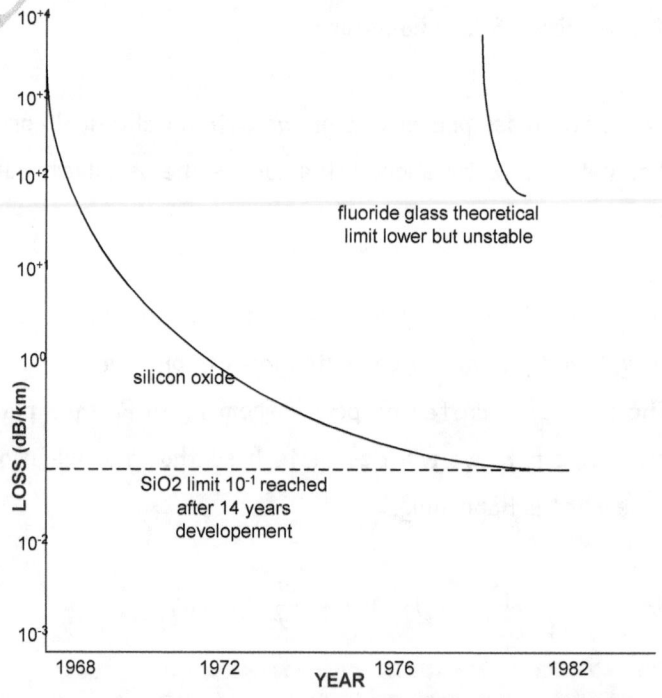

Figure 17. Attenuation (loss) for oxide and fluoride glass.

Summary 4

Dispersion in fibres is analysed generally but especially for the modal effects and the range of propagation parameters was found:

$$n_2 k \quad < \quad \beta \quad < \quad n_1 k$$

These limits are useful to compute number of guided modes for SI and GI profiles. The SI case was further studied to produce the eigenvalue equation for the β associated with each particular $LP_{\nu\mu}$ mode via Eq (44) along with Eq(40). Then the special case of SI monomode propagation was considered and the V-parameter below which that happens was derived using known cylinder function relations. Advanced profiles for WDM systems were discussed and polarisation maintaining designs were reviewed.

Answers to Ex 10-19

Ans Ex 10

Using given values: $n_1 - n_2 = 1.51 - 1.50 = 0.01$

$r = 25 \ \mu m$ $\lambda = 1.55 \ mm$

$r/\lambda = 16$ approx so index variation over 1λ is $0.01/16 = 0.0006$

This is negligeable in the context of $n_1 = 1.5$ the assumption in valid.

Ans Ex 11

$$\psi = F(r)\cos(\nu \phi)e^{i(\omega t - \beta z)}$$

$$\frac{d\psi}{dr} = \frac{dF(r)}{dr}\cos\nu\phi \exp i(\omega t - \beta z)$$

$$\frac{\delta^2\psi}{\delta r^2} = \frac{d^2F(r)}{dr^2}\cos\nu\phi \exp i(\omega t - \beta z)$$

$$\frac{\delta^2\psi}{\delta \phi^2} = F(r)\nu^2\cos\nu\phi \exp i(\omega t - \beta z)$$

$$\frac{\delta^2\psi}{\delta z^2} = (i\beta)^2 F(r)\cos\nu\phi \exp i(\omega t - \beta z)$$

Adding all according to cylindrical wave Eq(2) and cancelling   term $\cos\nu\phi \exp i(\omega t - \beta z)$ across gives us using on right hand side $k = 2\pi/\lambda$ with $\omega = 2\pi/\nu$ and from end of last section 2 that $c^2 = 1/\mu_0\varepsilon_0$:

$$\frac{\delta^2\psi}{\delta r^2} + \frac{1}{r}\frac{\delta\psi}{\delta r} + \frac{1}{r^2}\frac{\delta^2\psi}{\delta\phi^2} + \frac{\delta^2\psi}{\delta z^2}$$

$$= n^2\varepsilon_0\mu_0\frac{\delta^2\psi}{\delta t^2} \tag{2}$$

$$\Rightarrow \frac{d^2F}{dr^2} + \frac{1}{r}\frac{dF}{dr} + \left(n^2k^2 - \beta^2 - \frac{\nu^2}{r^2}\right)F = 0 \tag{3}$$

Ans Ex 12

$$\frac{d^2F}{dr^2} + \frac{1}{r}\frac{dF}{dr} + \left(n^2k^2 - \beta^2 - \frac{v^2}{r^2}\right)F = 0 \qquad (3)$$

Transform from linear r-space to log x-space using r = ae^x

Hence dr = ae^x dx = rdx and dx = (1/r)dr

Eq(3) times r² is starting step:

$$\Rightarrow \quad r^2\frac{d^2F}{dr^2} + r\frac{dF}{dr} + \left(k^2r^2 - v^2\right)F = 0 \text{ where } k^2 = n^2k^2 - \beta^2 \qquad (a)$$

$$dF = \frac{dF}{dx}dx = \frac{dF}{dx}\frac{1}{r}dr$$

$$\therefore \frac{dF}{dr} = \frac{1}{r}\frac{dF}{dx}$$

$$\frac{d^2F}{dr^2} = \frac{-1}{r^2}\frac{dF}{dx} + \frac{1}{r}\frac{d}{dr}(\frac{dF}{dx}) = \frac{-1}{r^2}\frac{dF}{dx} + \frac{1}{r^2}\frac{d^2F}{dx^2}$$

$$r^2\frac{d^2F}{dr^2} = -\frac{dF}{dx} + \frac{d^2F}{dx^2}$$

$$\therefore r^2(\frac{d^2F}{dr^2}) + r\frac{dF}{dr} = \frac{d^2F}{dx^2}$$

Hence Eq(a) above becomes:

$$\Rightarrow \quad \frac{d^2F}{dx^2} + \left(k^2r^2 - v^2\right)F = 0$$

$$\Rightarrow \quad \frac{d^2F}{dx^2} + \left(k^2a^2e^{2x} - v^2\right)F = 0$$

Ans Ex 13

Since $F(r) = A(r)e^{is(r)}$ **describes the radial amplitude and phase properties:**

$$\psi = F(r)\cos(\nu\phi)e^{i(\omega t - \beta z)} = A(r)e^{is(r)}\cos(\nu\phi)e^{i(\omega t - \beta z)} \qquad \text{MODE}$$

Now use exponential to trigonometric conversion $\cos\nu\phi = \frac{1}{2}(e^{i\nu\phi} \ e^{-i\nu\phi})$**:**

$$\therefore \psi = \frac{1}{2}A(r)\{\exp[i(\omega t - \beta z - \nu\phi + S(r))] + \exp[i(\omega t - \beta z + \nu\phi + S(r))]\} \qquad (13)$$

Ans Ex 14

Recall Eq(11)

$$S(r) = \int_1^r \left[\left(n^2 k^2 - \beta^2\right) - \frac{\nu^2}{r^2}\right]^{\frac{1}{2}} dr \qquad (11)$$

However we only used the positive square root term on first visit therefore for a complete solution the negative S(r) version needs to be added in:

$$\therefore \psi = \frac{1}{2}A(r)\{\exp[i(\omega t - \beta z - \nu\phi + S(r))] + \exp[i(\omega t - \beta z + \nu\phi + S(r))]\}$$
$$+ \frac{1}{2}A(r)\{\exp[i(\omega t - \beta z - \nu\phi - S(r))] + \exp[i(\omega t - \beta z + \nu\phi - S(r))]\}$$

Finally revert to trig from exp with a reverse conversion to tidy up:

$\cos\nu\phi = \frac{1}{2}(e^{i\nu\phi} + e^{-i\nu\phi})$ **and** $\cos S(r) = \frac{1}{2}(e^{iS(r)} + e^{-iS(r)})$ **:**

$$\therefore \psi = A(r)e^{is(r)}\cos(\nu\phi)\cos[S(r)]e^{i(\omega t - \beta z)} \qquad \mathbf{(14)}$$

Ans Ex 15

Eq(19) $\left[\Delta\tau_{opt} \approx \dfrac{Ln_1}{8c}\Delta^2\right]$ for optimised-profile modal dispersion with typical

fibre values gives using 1 km span:

$$\left[\Delta\tau_{opt} \approx \dfrac{Ln_1}{8c}\Delta^2 = 1.5(\dfrac{1.5-1.49}{1.5})^2\dfrac{1}{(24)10^8}\right]$$

This is 0.03 ps/km for GI parabolic index profile which is hundreds of thousands times improvement on SI fibre. Hence the deep cusp at $\gamma = 2$.

Ans Ex 16

Cut-off max and min values for β are found by letting β get so large it pulls the plot in Figure (7) entirely below the axis. The expression $n_1^2(r)k^2-\beta^2 = 0$ as the index peak crosses the axis where $n(r)$ is n_1.

Hence $\beta_{max} = n_1k$

In the reverse situation the plateaux for cladding where $n(r) = n_2$ passes the axis as b is so small giving $n_2^2(r)k^2-\beta^2 = 0$

Hence $\beta_{min} = n_2k$

Ans Ex 17

Number of guided modes.

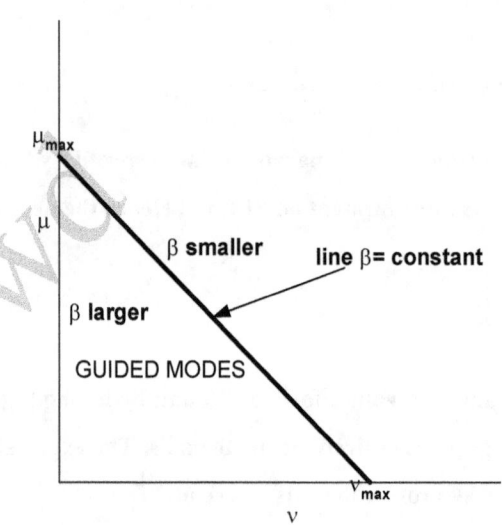

Figure 8   Guided LP modes depicted in mode number space

Since each LP mode is defined by two integers we can draw a 2D chart, Figure 8, which depicts each mode as a dot with its ν number on the horizontal axis and μ number on the vertical. There turns out to be a line above which the mode does not exist as the integers rise above an allowed limit for propagation constant β. Recall we have already found the permissible range of β.

When we now count the dots under that line the result is the number of modes in the guide.

Observe that the count is essentially the same as the area under the curve which we get by integration. Hence the number M of modes, with β in the permitted range, equals the integral of vertical μ with respect to horizontal $d\nu$:

$$
\begin{aligned}
M(\beta) &\approx 4\int_0^{\nu_{max}} \mu \, d\nu \\
&= \frac{4}{\pi}\int_0^{\nu_{max}} \int_{r_1}^{r_2} \frac{1}{r}\left\{\left[n^2(r)k^2 - \beta^2\right]r^2 - \nu^2\right\}^{\frac{1}{2}} dr \, d\nu
\end{aligned}
\tag{22}
$$

Here we substituted the expression for μ from the WKB analysis; that is the inner integral divided by π (strictly that gives $\mu+1/2$ but we assume μ is large).

But for completeness we have included a *multiplier 4* since there are both sine and cosine possibilities that are equally valid or periodic at the starting point Eq (2) and furthermore there can be two polarisation states of the same mode. So the single integral in Eq(22) is immediately expanded to double integral by substituting for μ using the WKB eigenvalue Eq (16)

$$
S(r) = \int_{r_1}^{r_2}\left[n^2(r)k^2 - \beta^2 - \frac{\nu^2}{r^2}\right]^{\frac{1}{2}} dr = \left(\mu + \frac{1}{2}\right)\pi = \mu\pi
\tag{16}
$$

Observe that in the Figure above ν is max when μ is zero and r_1 goes to zero because that is the LP mode with no radial repeats in the pattern.

Also as the rhs is now zero in Eq(16) the square bracket on lhs is zero with β minimised at $\beta_{min} = n_2 k$

$$
\left[n^2(r)k^2 - \beta_{min}^2 - \frac{\nu_{max}^2}{r^2}\right] = 0
$$

Hence:

$$
\frac{\nu_{max}^2}{r^2} = n^2(r)k^2 - \beta_{min}^2 = n^2(r)k^2 - n_2^2 k^2
\tag{A}
$$

Profile:
$$n(r) = \left[1 - 2\Delta\left(\frac{r}{a}\right)^{\gamma}\right]^{\frac{1}{2}} \quad \cdots \cdots \quad r < a$$
$$n_2 = n_1\left[1 - 2\Delta\right]^{\frac{1}{2}} \quad \cdots \cdots \quad r \geq a$$

(17)

Hence
$$n^2(r) - n_2^2 = \left[2n_1^2\Delta\left(1 - \frac{r}{a}\right)^{\gamma}\right]$$

B

Sub this Eq B into A above gives

$$v_{max}^2 = 2r^2 n_1^2 k^2 \Delta\left[\left(1 - \frac{r}{a}\right)^{\gamma}\right]$$

C

In Eq(22) as the range is now 0 to $v_{\mu\alpha\xi}$ and r goes 0 to a to capture all modes possible let us switch to N for mode number.

Using $\quad n(R)k = \beta$ and keeping in mind the range for β given by Eq (21), namely $\quad n_2 k \quad < \quad \beta \quad < \quad n_1 k \quad$, we get using Eq A above:

$$N = \frac{4}{\pi}\int_0^{max}\int_0^a\frac{1}{r}\left\{\left[n^2(r)k^2 - \beta_{min}^2\right]r^2 - v^2\right\}^{\frac{1}{2}}drdv \qquad (23)$$

Reverse order of integration:

$$N = \frac{4}{\pi}\int_0^a\frac{1}{r}\int_0^{max}\left[v_{max}^2 - v^2\right]^{\frac{1}{2}}dvdr \qquad (24)$$

The integral with respect to v is well known as an arcsine functional relation in the mathematical handbook (or using integration by parts) producing a $\pi/4$ that then cancels by good fortune the $4/\pi$ with the result:

$$N = \frac{4}{\pi}\int_0^a\frac{1}{r}\left[\frac{\pi v_{max}^2}{4}\right]dr = \int_0^a\frac{1}{r}\left[v_{max}^2\right]dr$$

But according to C: $\quad v_{max}^2 = 2r^2 n_1^2 k^2 \Delta\left[\left(1 - \frac{r}{a}\right)^{\gamma}\right]$

$$N = \int_0^a\frac{1}{r}2r^2 n_1^2 k^2 \Delta\left[\left(1 - \frac{r}{a}\right)^{\gamma}\right]dr$$

Thus we get an integrand a sole function of r. Here we must bring r inside the bracket so the , so by Eq (24) the exponent becomes γ+1, and then integrated so that the power becomes γ+2:

$$N = 2n_1^2 k^2 \Delta \left[\left(\frac{r^2}{2} \right) - \frac{1}{a^\gamma} \left(\frac{r^{g+2}}{\gamma^2} \right) \right]$$ and this definte integral must be

evaluated between limi.. r=0 to r=a

giving $$N = 2n_1^2 k^2 \Delta a^2 \left[\frac{1}{2} - \frac{1}{\gamma + 2} \right]$$

Finally: $$N = \left[\frac{\gamma}{(\gamma + 2)} \right] (n_1 ka)^2 \Delta \qquad (28)$$

Here we have renamed the general M(β) simply as N, the number of modes below the line for the limiting value of β and therefore the total number.

Let us for convenience define the fibre V-parameter: $\left(V \equiv n_1 ka \sqrt{2\Delta} \right)$

This along with eq. (28) $\Rightarrow$ $$N = \frac{\gamma}{2(\gamma + 2)} V^2 \qquad (29)$$

Ans Ex 18

$$\kappa = \left(n_1^2 k^2 - \beta^2\right)^{\frac{1}{2}} \tag{35}$$

$$\gamma = \left(\beta^2 - n_2^2 k^2\right)^{\frac{1}{2}} \tag{36}$$

$$\left[(\kappa a)^2 + (\gamma a)^2 = V^2\right] \tag{40}$$

$$\left[\kappa J_{v+1}(\kappa.a)K_v(\gamma a) = \gamma J_v(\kappa.a)K_{r+1}(\gamma a)\right] \tag{44}$$

Eq (40) and (44) have variables κa and γa but selected parameters V and a depending on the fibre chosen. Once κ and γ are found then they are inserted into Eq (35) and (36) to quantify β.

Ans Ex 19

$\left(V \equiv n_1 ka\sqrt{2\Delta}\right)$; $\Delta = (n_1 - n_2)/n_1 = 0.01/1.5$ and $k = 2\pi / \lambda$

Hence a = $2.405/2\pi(0.115)$ in microns

So a= 3.33 μm for the single mode fibre or diameter D= 6.6 μm.

This means single mode fibre is about 10 times thinner core than multimode but also the splicing and connectors must be ten times more precise. So we need top quality mechanical engineering and laser monitoring. We will need to use OTDRs (optical time domain reflectometers) to assess the precision of all joints.

Conclusion

We have discovered by mathematics how to create broadband over fibre.

Dispersion in its many guises was scientifically modelled to produce optimal refractive index profile design, then implemented by computer controlled chemical vapour deposition. May you enjoy the broadband communications that results from your new understanding of these "glassworks" of wonderful engineering.

Typical Examination Questions.

$$c = 3 \times 10^8 \quad ms^{-1}$$
$$e = 1.6 \times 10^{-19} C$$
$$h = 6.6 \times 10^{-34} Js$$
$$k = 1.38 \times 10^{-23} JK^{-1}$$
$$m_e = 9.11 \times 10^{-31} kg$$

Note: **Bessel and Hankel function relations are provided below at end paper.**

QUESTION A

(a) By considering *all* solutions to the following wave equation

$$\nabla^2 \psi = n^2 \varepsilon_0 \mu_0 \frac{\delta^2 \psi}{\delta t^2}$$
(1)

show that the *complete* expression for the guided mode is as equation (2) and that the corresponding eigenvalue equation for guided light in an optical fibre is given by equation (3):

$$\psi = A(r)\cos(v\phi)\cos[S(r)]e^{i(\omega t - \beta z)}$$
(2)

$$\int_{r_1}^{r_2} \left[n^2(r)k^2 - \beta^2 - \frac{v^2}{r^2} \right]^{\frac{1}{2}} dr = \left(\mu + \frac{1}{2} \right)\pi$$
(3)

50%

(b) Indicate briefly what further modification is taken into account before assessing the number of guided modes.

10%

(c) Having now solved for the guided modes state why equation (1) is so called.

This result contains the refractive index profile n(r). Outline how you would use computer methods to apply your result and discover numerically the optimum profile for multimode fibres and plot the expected outcome.

20%

(d) Outline how you would apply this result in an engineering context to manufacture fibres of the desired n(r) design.

20%

QUESTION B

Consider a general optical fibre index profile:

$$n(r) = n_1 \left[1 - 2\Delta \left(\frac{r}{a} \right)^s \right]^{\frac{1}{2}}$$

where s is the "shape factor" and other symbols have the usual meanings.

Given the eigenvalue equation for the LP modes:

$$\kappa J_{\nu+1}(\kappa a) K_\nu(\gamma a) = \gamma J_\nu(\kappa a) K_{\nu+1}(\gamma a)$$

where κ and γ depend on propagation constants and are related to the fibre V-parameter, show that the condition for single mode operation (i.e. cut-off condition) is:

$$V_C = 2.405$$

At what point in this argument is s value of infinity assumed?

Hence indicate briefly how the condition:

$$V_C = 4$$

might be arrived at as the corresponding condition for graded index fibre. Illustrate the trend for other profiles and discuss how and where this might find application in optical communication systems.

QUESTION C

(i)   Evaluate the maximum and minimum modal propagation constants β, in terms of free space wavelength and optical fibre parameters. Describe the mode cut-off mechanisms and explain the term "leaky mode".

(ii)   What is meant by "cut-off wavelength" and describe how it can be measured. Describe how this is related to core dimension or other fibre properties.

QUESTION D

Show that in a typical optical fibre refractive index profile design for telecommunications the basic starting assumption of the WKB theory will apply.

Establish the phase conditions for the guided modes (multimode case) and hence the **"turning point"** condition. Describe what happens in the various regimes for the modal propagation constant implied by this condition.

QUESTION E

(a)   Consider a general optical fibre index profile:

$$n(r) = n_1 \left[1 - 2\Delta \left(\frac{r}{a} \right)^s \right]^{\frac{1}{2}}$$

where s is the "shape factor" and other symbols have the usual meanings.

Given the eigenvalue equation for the LP modes:

$$\kappa J_{v+1}(\kappa a) K_v(\gamma a) = \gamma J_v(\kappa a) K_{v+1}(\gamma a)$$

Where κ and γ depend on propagation constants and are related to the fibre V-parameter, show that the condition for single mode operation (i.e. cut-off condition) is:

$$V_C = 2.405$$

(b)   At what point in the above argument is s = ∞ assumed?   Hence indicate briefly how the condition

$$V_C = 4$$

might be arrived at as the corresponding condition for graded index fibre. Illustrate the trend for other profiles and discuss how and where this might find application in optical communication systems.

QUESTION F

Use the WKB theory to find the number of guided modes in an optical fibre with refractive index profile

$$n(r) = n_1 \left[1 - 2\Delta \left(\frac{r}{a} \right)^s \right]^{\frac{1}{2}}$$

where s is the "shape factor" and other symbols have the usual meanings.

Show that parabolic index fibre carries half the light of a step index fibre.

QUESTION G

(i)   Dispersion-shifted optical fibre has zero total dispersion at the loss minimum for the glass. Discuss how this arises and how it is achieved for the combined dispersion contributions. Describe further advanced refractive index profiles and their purpose.

(ii) The loss mechanisms in the fibre produce the ultimate minimum in mid C-band. Describe these and their respective spectra.

QUESTION H

A general index profile optical fibre is analysed by the WKB method and produces the expression

$$M(\beta_{min}) = \frac{4}{\pi} \int_{r=0}^{a} \frac{1}{r} \left[\frac{\pi \, v_{max}^2}{4} \right] dr$$

for the number of guided modes, the symbols taking their normal meaning. Use this along with the Gloge and Marcatili expression for the index profile to show that approximately half as much energy may be carried by a graded-index fibre as a step-index fibre of corresponding dimensions.

QUESTION I

(i)   For the case of a multimode fibre the $LP_{\upsilon\mu}$ modes $F(r)$ satisfy:

$$\frac{d^2 F}{dr^2} + \frac{1}{r}\frac{dF}{dr} + \left[n^2(r)k^2 - \beta^2 - \frac{\upsilon^2}{r^2} \right] F = 0$$

The symbols take their usual meanings. By defining appropriate parameters κ and γ for the core and cladding regions respectively and then relating these to the V-parameter for the fibre, show that an eigenvalue equation for the LP modes can yield the modal propagation constants.

(ii)   Use the same eigenvalue equation to prove that in a step-index single-mode fibre the V-parameter associated with cut-off must be zero in order that the LP_{01} mode be leaky.

State the tendency towards this condition in terms of:

(a)   fibre radius,
(b)   transmission wavelength,
(c)   index profile.

  Note: Bessel and Hankel function relations are provided below.

Mathematical properties used.

Exponential to trigonometric conversion:

$e^{iv\phi} = \cos v\phi + i\sin v\phi$

$e^{-iv\phi} = \cos v\phi - i\sin v\phi$

$\cos v\phi = \tfrac{1}{2}(e^{iv\phi} + e^{-iv\phi})$

Cylinder functions:

The Bessel function $J_v(Kr)$ remains finite at the co-ordinate origin $r = 0$ and is therefore the appropriate cylinder function solution to eq (32) for the fibre core region:

$$F(r) = AJ_v(\kappa.r) \quad \cdots \quad |r| < a \tag{34}$$

The modified Hankel function $K_v(\gamma r)$ is the cylinder function with imaginary argument and vanishes as $r \to \infty$. and applies to the cladding.

Functional relations for Hankel functions:

$$\left.\begin{array}{lll} I & K_v(x) = & \ln\left(\dfrac{2}{x}\right) \quad \cdots \; v = 0 \\[2mm] \textit{and } II & K_v(x) = & \dfrac{(v-1)!}{2}\left(\dfrac{2}{x}\right)^2 \quad \cdots \; v \geq 1 \end{array}\right\} \tag{48}$$

Functional relation for three orders of Bessel function:

$$V_c J_{v-1}(V_c) + V_c J_{v+1}(V_c) = 2vJ_v(V_c) \tag{52}$$

For the zero order Bessel function to be zero $J_0(V_c) = 0$ the argument is determined as the plot crosses the axis at 2.405:

$$\underline{V_c = 2.405} \tag{54}$$

Hardware: OptoSci Kit

See OptoSci Manual and proceed with the selected hardware or HW experiments below.

HW-1
System losses and bandwidth.

Experiments 8.1, 8.2, 8.3 pp12-15 ED-COM Manual and B2 Measure fibre length by pulse transmission (OptoSci Appendix B in the ED-COM Manual).

HW-2
Eye Diagrams and bit-error rate BER

See Fig 2.3, Eq (13), Fig 2.7,

Experiment 6.1   BER(COM) OptoSci Manual pp 21
          6.2   22
          6.3   23
          6.4   24
          6.5   25

Name: _____ Student Number: _____

1.1 Draw the characteristic of output power against drive current by hand for LED and LASER. Then, mark the important points (bias point, threshold current).

LED

I [mA]								
P [uW]								

LASER

I [mA]								
P [uW]								

Graph:

2.1 Measure the output value for the patchcord (end-to-end), reel #1 and reel #2 for LED and LASER. Then, calculate the attenuation of link and compare. Make a comparison and evaluation.

Link/Source	P [uW]

Other parameters:

2.2 Calculate the length of fibre for each reel from the time difference value of biased signal and detecting signal after the fibre link. Do it for LED and LASER source. Consider results.

3.1 Modulate the LED source by square signal. Then, find the pulse risetime (10% to 90%) on the oscilloscope. Take measurements of both reel, interconnected reels and end-to-end scheme. Consider system and fibre properties. Compare the values Bandwidth-length product and bitrate-distance product.

Setup	U_{10} [mV]	U_{90} [mV]

Other parameters:

3.2 Repeat measurement of Bandwidth-length of interconnected reels and end-to-end setup. Modulate LED in the range of 2 MHz to 28 MHz. Obtain and draw by hand results for fibre frequency response

Setup: interconnected

f [MHz]	U_{out} [mV]

Other parameters:

Setup: end-to-end

f [MHz]	U_{out} [mV]

Other parameters:

Name: _____   Student Number: _____

HW-2.   Eye Diagrams and BER

See Fig 2.3, Eq (13), Fig 2.6, 2.7,

Experiment 6.1   BER(COM) Manual pp   21
          6.6   22
          6.7   23
          6.8   24
          6.9   25
          Results table   44

6.1 Set up pseudo-random bit-sequence PRBS generator at 10,20 and 40
     Mbit/s
     CRO at 2 V/div or 1 V/div (uncal) and 5 ns/div
     Observe eye-diagram
     Measure tr and tf at each bit-rate
     With cursor on, or by visual inspection expect about 10 ns at 40
     Mbit/s for example.

6.2 LED TX at 50 mA and 6 microwatt optical at patchcord
     50 ohm CRO termination to avoid reflections (dips)
     Measure tr, tf (manual p21) Expect about 11.5 ns or 45% at 40
     Mbit/s

     At 80% of v1measure t-pulse expecting about 16.5 ns or about 64%
     of bit-period

     Measure jitter on eye expecting about 500 ps

6.3 Use template p28 for results
Measure tr and t-pulse for 1,2 and 3 km cable
Example: at 3 km + patchcord t-pulse of about 23.5 ns

At 50% level measure full jitter; expect about 1ns after 1 km and 2 ns after 3 km
This due to amplitude noise converting to time-spread on sloping rise/fall

6.4 At 1,2 and 3 km measure eye Q factor or (v1-vt)/vN
where vN is noise spread of v1 at 20 mV/div
Expect about Q = 6 for 3 km

Use Fig to estimate BER from Q^2
Expect about 0.8×10^{-9}

Measure jitter; expect about 2 ns due to amplitude noise converting to time spread
(Was about $\frac{1}{2}$ ns at 6.2)

6.5 Repeat 6.2 to 6. 4 with laser TX at 33 mA, 10 to 40 Mbit/s, about 520 microwatt

Then insert 1 to 3 km cables and tabulate per template

Expect tr about 10.5 ns; and 15.5 ns for 3 km

Measure jitter (expect about 1.5 ns) and Q
Then estimate BER   Expect Q about 7.5 giving BER 10^{-12}

Tabulate results.

INDEX

Ronan O'Dowd © 2011
(100 pages)

Do not copy electronically or in print.

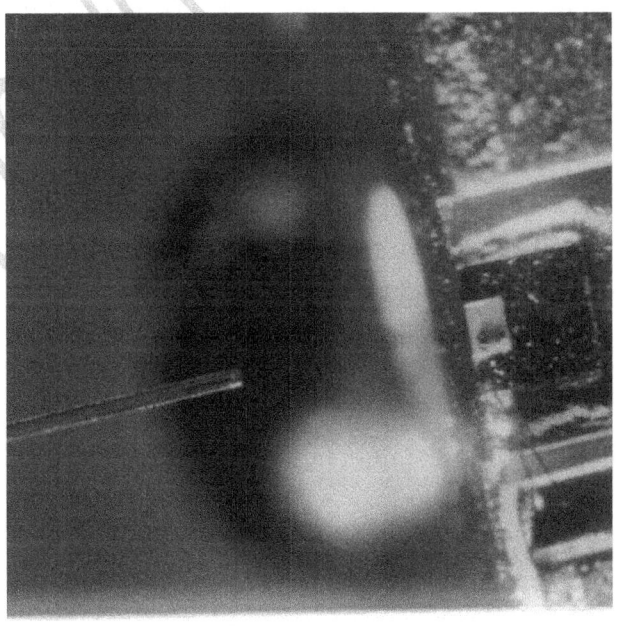

Ronan O'Dowd's Designer Books in Photonic Engineering:

Photonics Handbook Part 2

"Photonics Handbook Part 2: LEDs, Lasers, Detectors"

"Photonics Handbook Part 2: LEDs, Lasers, Detectors"
Author

Ronan O'Dowd PhD SMIEEE is Professor Emeritus Photonic Engineering at UCD Dublin, Ireland where he taught and researched Optoelectronics and Photonics for three decades until 2010. He has several breakthrough papers in topics such as tunable semiconductor lasers and optical communications, including the millennium 2001 paper proving a dense comb of 2000 wavelength channels could be transmitted in a single fibre using the same semiconductor laser (ref *IEEE Jnl.S.T. Quantum Electronics Mar 2001*). Many of his students have proceeded to successful careers in academia and the photonics industry worldwide.

By the same author: Physics Science of Action   **Gill and Macmillan 1984**

"Photonics Handbook Part 2: LEDs, Lasers, Detectors"
Tips to use this guidebook.

This series of books for photonic system designers will cover the subsystems that make up an optical communications link. These are the transmitter, fibre channel and receiver and since the fibre design sets the specifications and key criteria to be implemented at either end we tackled that in Part 1. Part 2 covers the terminals

The guide is formatted with *You Do* exercises and answers are provided. These are intended to be part of the learning process that will take the student to Engineering degree high level over what may be about a 30 hour degree module where 6 additional hours are set aside for practical work. The hardware link kit from OptoSci, where optical fibre dispersion etc can be measured over fibre reels, is very useful for laboratory explorations.

This Part 2 now covers laser, detector and system design for broadband and may constitute a further module.

COVER PICTURE: An optical fibre 1/8 mm thick faces a semiconductor laser chip seated on a metal sub-mount in the author's laboratory.

CONTENTS PART 2

Notes:

You Do exercises should be attempted especially by the self-taught student using this guide and regardless of your confidence in your answer quality. Having then read the answer provided you should again attempt it.

Diagrams are simple line-style that the student should re-draw.

There are sample examination questions at the end using standard mathematical relations.

1 Materials for Photonic Devices

The commonest semiconductor for electronics is silicon, Si, and is in group 4 (IV) of the periodic table of elements (Table 1 and Appendix 1). We will see that it can also be used, but only in restricted circumstances, for *photonics*, where light particles carry the information in place of electrons. We must therefore seek other materials that are more suited to the emission and detection of light and that is the realm of photonics. We will find that a family of semiconductor alloys consisting of two (binary), three (ternary) or four (quaternary) elements from groups 3 (III) and 5 (V) of the periodic table will satisfy our requirements adequately.

To understand these alloys we must review briefly the band theory of semiconductors as it applies to group IV silicon and group III-V alloys such as gallium arsenide GaAs and indium phosphide InP.

Table 1 Periodic Table of Elements

Group:

I	II	3	4	5	6	7	8	9	10	11	12	III	IV	V	VI	VII	VIII
1																	18
1 H 1.008	2											13	14	15	16	17	2 He 4.0026
3 Li 6.94	4 Be 9.0122											5 B 10.81	6 C 12.011	7 N 14.007	8 O 15.999	9 F 18.998	10 Ne 20.180
11 Na 22.990	12 Mg 24.305	3	4	5	6	7	8	9	10	11	12	13 Al 26.982	14 Si 28.085	15 P 30.974	16 S 32.06	17 Cl 35.45	18 Ar 39.948
19 K 39.098	20 Ca 40.078	21 Sc 44.956	22 Ti 47.867	23 V 50.942	24 Cr 51.996	25 Mn 54.938	26 Fe 55.845	27 Co 58.933	28 Ni 58.693	29 Cu 63.546	30 Zn 65.38	31 Ga 69.723	32 Ge 72.63	33 As 74.922	34 Se 78.96	35 Br 79.904	36 Kr 83.798
37 Rb 85.468	38 Sr 87.62	39 Y 88.906	40 Zr 91.224	41 Nb 92.906	42 Mo 95.96	43 Tc (98)	44 Ru 101.07	45 Rh 102.91	46 Pd 106.42	47 Ag 107.87	48 Cd 112.41	49 In 114.82	50 Sn 118.71	51 Sb 121.76	52 Te 127.60	53 I 126.90	54 Xe 131.29
55 Cs 132.91	56 Ba 137.33	57-71 *	72 Hf 178.49	73 Ta 180.95	74 W 183.84	75 Re 186.21	76 Os 190.23	77 Ir 192.22	78 Pt 195.08	79 Au 196.97	80 Hg 200.59	81 Tl 204.38	82 Pb 207.2	83 Bi 208.98	84 Po (209)	85 At (210)	86 Rn (222)
87 Fr (223)	88 Ra (226)	89-103 #	104 Rf (265)	105 Db (268)	106 Sg (271)	107 Bh (270)	108 Hs (277)	109 Mt (276)	110 Ds (281)	111 Rg (280)	112 Cn (285)	113 Uut (284)	114 Uuq (289)	115 Uup (288)	116 Uuh (293)	117 Uus (294)	118 Uuo (294)

* Lanthanide series	57 La 138.91	58 Ce 140.12	59 Pr 140.91	60 Nd 144.24	61 Pm (145)	62 Sm 150.36	63 Eu 151.96	64 Gd 157.25	65 Tb 158.93	66 Dy 162.50	67 Ho 164.93	68 Er 167.26	69 Tm 168.93	70 Yb 173.05	71 Lu 174.97
# Actinide series	89 Ac (227)	90 Th 232.04	91 Pa 231.04	92 U 238.03	93 Np (237)	94 Pu (244)	95 Am (243)	96 Cm (247)	97 Bk (247)	98 Cf (251)	99 Es (252)	100 Fm (257)	101 Md (258)	102 No (259)	103 Lr (262)

You do...

Ex 1: Elements for photonics.

Locate the elements silicon Si, gallium Ga, arsenic As, indium In, and phosphorous P by period and group in the Periodic Table of elements Table 1 / Appendix 1. Draw a sub-set of the main table three groups wide and two periods deep containing all of these elements.

Answer Ex 1 at end handbook and in Appendix 1.

○ End of exercise

The energy band diagrams of Si and GaAs, both being semiconductors, consist of a conduction band separated by a forbidden gap E_g from the valence band, Figure 1.

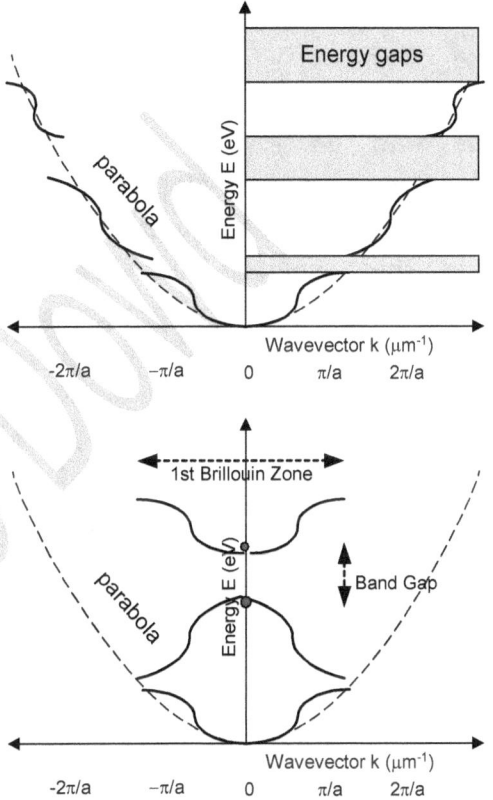

Figure 1. Band diagrams show E (energy) versus k (wavevector) for photonic semiconductors. The top two bands are then isolated and segments translated to the +/- π/a range of k to produce the reduced Brillouin zone representation.

These diagrams are derived in semiconductor coursework by applying the Schroedinger equation in the context of the periodic potential $V(x)$ experienced by the electron waves in a crystal of periodicity a. In the one dimensional or 1D case $V(x) = V(x+a) = V(x+2a) = $ ...etc. The outcome of that analysis is the allowed energies showing discontinuities along the electronic wave vector axis k wherever $k = +/- n\pi/a$ (Figure 1 upper). The top two bands are then isolated and segments translated to the $+/- \pi/a$ range of k to produce the reduced *Brillouin zone* representation (Figure 1 lower).

The upper two levels alone are of interest to us here since lower bands require X-ray energies to access them. Visible and infra-red or IR energy light can access the two upper bands, called conduction and valence, that are separated by a forbidden gap E_g. That gap is about 1 electron volt, 1 eV, depending on the material.

There is a crucial distinction between the E-k diagram for Si, Group IV, and that for GaAs or the other III-V alloys. With GaAs (Figure 2 upper) the bottom of the conduction band resides directly above the top of the valence band so it is called a *direct bandgap semiconductor*. For Si however this is not the case (Figure 2 lower) so it is called an *indirect bandgap semiconductor*. This means that along with an energy gap E_g between these positions on the diagram there is also a horizontal gap along the k-axis where wave-vector k is related to and a measure of the momentum p for the electron.

Recall: electron wave-vector k = 2π/λ while momentum p = h/λ. Therefore p is proportional to k and the E-k band diagram effectively plots electron energy versus momentum.

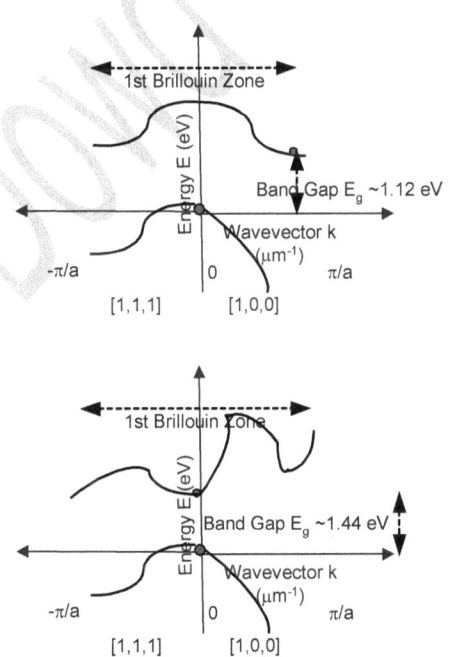

Upper represents Si indirect band gap 1.12 eV; lower GaAs direct bandgap 1.44 eV

Figure 2. Band diagrams for GaAs and Si showing added detail with crystal directions [1,1,1] at left of E-axis and [1,0,0] at right of E-axis. (This is done for data compression, otherwise left is just a reflection of right).

An electron in gallium arsenide that recombines with a hole must transit vertically releasing energy equal to E_g and that can be in the form of a photon of light. (See Ex 2). For this to happen with silicon however there must be a momentum shift as well in order to move horizontally along the k-axis from bottom of conduction to top of valence bands (dots in Figure 2). To obey the law of conservation of momentum a third particle must participate that gives or takes away momentum. There is only a slight probability that a third particle exists nearby at this instant so the likelihood of a photon being created in this way is very small. *Injection luminescence*, the emission of light by inputting current, is not expected from silicon or indeed found except in novel, artificial forms of that element.

You do...
Ex 2: Wavelengths and energy gaps.

Calculate the wavelength associated with a 1 eV energy gap and also for Si and GaAs where it is 1.12 eV and 1.44 eV respectively.
Answer Ex 2 at end handbook.
   o End of exercise

   o

The 3D solution to the Schroedinger equation is more complicated but similar diagrams to 1D apply corresponding to different directions in the crystal. The 1st Brillouin zone has shape dependent on crystal structure. Boundaries are still close to π/a where a is defined by the unit cell dimension. In real crystals the maximum of the valence band does not always occur at the same k value as the minimum of the conduction band. In an indirect bandgap semiconductor they do not while in a direct bandgap semiconductor they do coincide.

For this reason Si is not used in photonics for light emitting diodes, *LEDs*. Si may be used to detect light however as long as the light wavelength is in the visible or near IR and shorter than 1100 nm or 1.1 μm, corresponding to the bandgap of silicon (Exercise 2).

III-V alloys are the prime choice for luminescent semiconductors in the form of LEDs. These may be binary, ternary or quaternary alloy mixes depending on the light wavelength we wish to create. See Table 2 and Figure 6 below. They will detect at those IR wavelengths also so are suitable for photodiodes too.

Other materials that are indirect bandgap can be used to create photodiode detectors to suit particular wavelengths such as the 1800 nm region where germanium, Ge, has a bandgap. Broadband optical fibre operates in the 1550 nm window so that ternaries and quaternaries are the choice for both transmitter TX and receiver RX.

Injection Luminescence

The more detailed band diagrams for Si and GaAs are shown in Figure 2 above. GaAs is grown in a similar fashion to Si by *chemical vapour deposition, CVD* using extremely pure reactant gases. The carrier gas valves are programmed to allow input of elemental dopants at precise concentrations to produce layers that are p-type and n-type. By these means a wafer of pn diodes results. The wafer is then cut into chips that are packaged to give a light emitting diode or LED.

When current is injected into a positively-biased pn-diode the movement of electrons and holes is well known. With heavy doping and under forward bias minority carriers are injected from both sides Figure 3. The excess electron concentration at distance x from the junction in the p material falls exponentially with x.

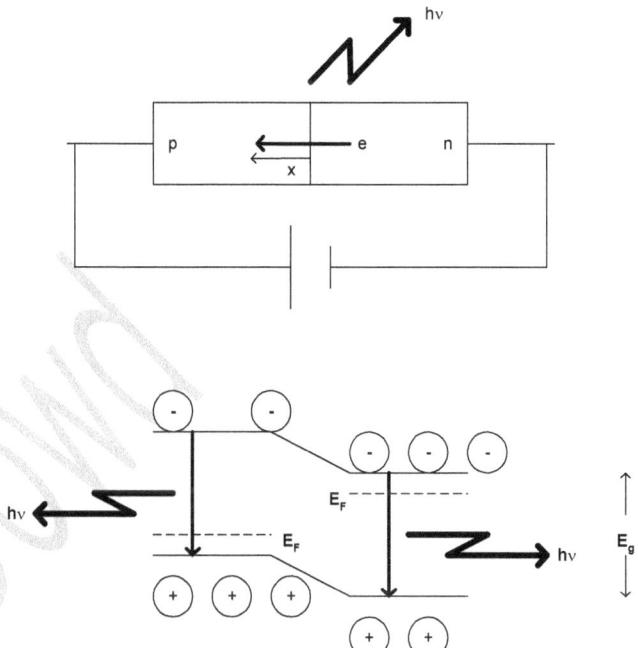

Figure 3. GaAs LED with forward bias showing carrier injection of electrons in the conduction and holes in the valence bands. Recombination after a penetration distance x produces IR light photons hν corresponding to the energy gap E_g.

Quantum efficiency is the rate of emission of photons divided by the rate of supply of electrons by injected current. The units are therefore watts of light per amp of current, WA^{-1}.

If *radiative recombination* does occur a photon is produced with wavelength λ_g given by Plank's equation:

$$\lambda_g = hc/E_g$$

Thermal energy ensures electrons in the conduction band have average energy kT/2 above the bottom of the band so that λ can be shorter than λ_g due to an increased jump. Meanwhile much recombination involves dopant energy levels within the gap so that λ may more likely be longer than λ_g producing a spread of "colours", the emission spectrum Figure 4. The above may be avoided when a *phonon* is available (a particle carrying the energy of a crystal vibration, essentially becoming heat) and then non-radiative recombination reduces the quantum efficiency.

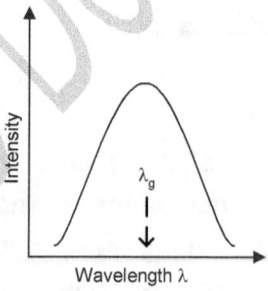

Figure 4. Emission spectrum for GaAs LED centred on λ_g.

You Do...
Ex 3: Optical power from LED.

A forward current of 30 mA is injected into a GaAs LED with quantum efficiency 95%. What output light power results? Use data from Ex 2.

Answer to Ex 3 at end handbook.

Interband Transitions

Transitions must conserve total wave-vector of the system. For a photon $k = 2\pi/\lambda$ while for an electron the range is $-\pi/a$ to $+\pi/a$ or $2\pi/a$ as shown in the Brillouin-zone diagram Figure 1. For 500 nm radiation (a visible LED) and typical lattice spacing 1 angstrom or 0.1 nm we find $2\pi/\lambda$ is thousands of times smaller than π/a. (Verify as an Exercise). Hence if only a photon and electron are involved then the transition is between states with virtually the same electron wave-vector (Figure 5a). If a phonon is also involved then a non-vertical transition may be possible (Figure 5b). In the latter case the phonon may contribute its energy E_p by annihilation or deduct E_p

by being created as a crystal vibration from the process:

$$hc/\lambda = E_g +/- E_p$$

Phonon energies are typically only 0.01 eV so that the light wavelength is approximately equal to λ_g while momentum is still conserved. However since three particles must be involved in the same place at the same time this transition probability is very low.

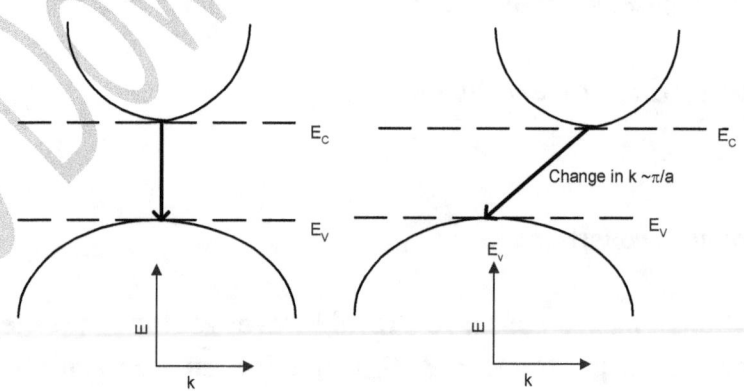

Figure 5. Direct and indirect transitions.

Interband *transition rate* r may be written:

$$r = Bnp$$

Here the participant carrier concentrations are n electrons and p holes while the coefficient B is assigned to each material. Note from Table 3 that for direct bandgap semiconductors B is approximately 10^5 times greater than indirect bandgap materials like Si. Unless other mechanisms are present, therefore, the latter, like silicon, are not suitable for LEDs.

TABLE 3   Common photonic materials

Material	Bandgap	E_g (eV)	B (m^3s^{-1})	λ_g (nm)
Si	indirect	1.12	1.79×10^{-21}	1106
Ge	indirect	0.67	5.25×10^{-20}	1880
GaP	indirect	2.26	5.37×10^{-20}	549
GaAs	direct	1.44	7.21×10^{-16}	861
InP	direct	1.35	1.26×10^{-15}	918
CdTe	direct	1.50	library task	826

Alloy Map for InGaAsP

Various mixes of the four elements indium In, gallium Ga, arsenic As, and phosphorous P produce injection luminescence at different wavelengths because the energy gap depends on the alloy chosen. The mole fraction for In is designated x so that for Ga must be (1-x) as one substitutes for the other in the crystal structure. Equally if As has mole fraction y then P must be at (1-y). The alloy is designated $In_xGa_{1-x}As_yP_{1-y}$ and the lattice parameter or atomic spacing in the crystal depends on x and y. Since the energy gap E_g is dependent on the lattice parameter a we may draw an alloy map that summarises the important photonic information, Figure 6.

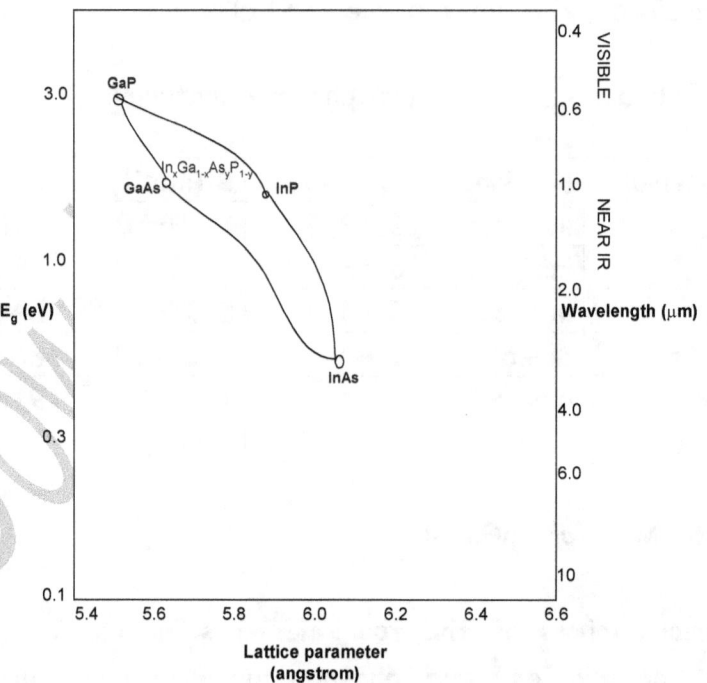

Figure 6. Alloy map for $In_xGa_{1-x}As_yP_{1-y}$ with ternary compounds along the perimeter and quaternaries inside.

The vertical axis can be both E_g and λ as $E_g=hc/\lambda$ and this is fundamental information to selecting the most suitable material for a given diode design. It is evident that we can select x and y to create visible or infra red LEDs and lasers by this means called *band-gap engineering*. A selection of relevant data for the binaries, these reside at corners of the map, is given in Table 4.

Table 4.   Physical properties of photonic materials

	Units	InP	GaAs	Si
Lattice parameter	angstrom 10^{-10} m	5.9	5.6	Library exercise
Band gap	eV	1.35	1.43	
Optical transition		Direct	Direct	
Electron mobility	cm^2/Vs	4500	8500	
Hole mobility	cm^2/Vs	150	420	

Exercise: Find in the library and fill into table 4 the relevant data for Si. Calculate and fill a row for λ_g also.

LED Construction and Efficiency

The map is used and a selected alloy is grown by liquid phase epitaxy LPE or vapour phase deposition VPD onto a wafer substrate and then diced into chips which are metallised for electrical contact, Figure 7.

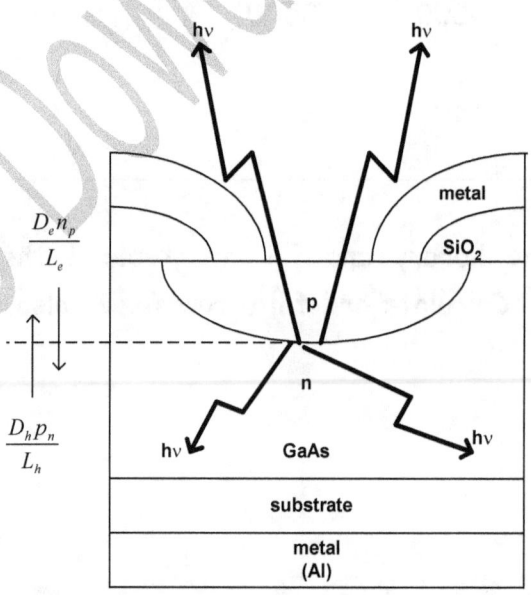

Figure 7. Surface emitting GaAs   LED for 850 nm wavelength in the near-IR with insulator SiO₂ and metallisation contacts.

This is a simple surface emitting LED whose active region, where the photons are created, is designed for

850 nm light in the near-IR. A disc window is etched into the surface layer to permit exit of the radiation through the silicate insulator SiO_2 that covers the upper p-layer of the pn structure. Measures of injected carriers $D_e n_p / L_e$ and $D_h p_n / L_h$ are shown that contribute to the diode current. D is the relevant diffusion coefficient and L the diffusion length for electrons or holes while n_p and p_n are electron and hole carrier concentrations in the p and n materials respectively.

Internal Quantum Efficiency

Radiative recombination must mainly occur on the side of the pn junction nearest the surface in order to lessen re-absorption. Hence we ensure that most forward current is due to carriers injected up into the surface p-layer of this "n-side down" device. The fraction q_e carried by electrons injected into the p-side is the internal quantum efficiency as it is these that contribute to emitted power:

$$q_e = (D_e n_p / L_e) / [(D_e n_p / L_e) + (D_h p_n / L_h)]$$

Dividing by the upper bracket gives

$$q_e = [1 + (D_h L_e p_n / D_e L_h n_p)]^{-1}$$

Now use the Einstein relation for electron or hole D with mobility μ namely $D_{e,h} = (kT/e)\mu_{e,h}$ and also the basic semiconductor equation

$$n_p p_p = n_n p_n = n_i^2$$

The result becomes

$$q_e = [1 + (\mu_h p_p L_e / \mu_e n_n L_h)]^{-1}$$

In III-V compounds electron mobility μ_e is largely dominant over holes μ_h while L_e and L_h are similar in value and cancel. Now assuming we design a n^+p device where the doping n_n greatly supersedes p_p the outcome is that the inner bracket is tiny. This means q_e is almost unity or quantum efficiency approaches 100%. Observe that all this derives in the case of an n-side down device.

Exercise: If it were to be p-side down you should follow the argument through again to find that q_h can be made close to 100% by using a p^+n doping regime.

Exercise: Re-draw Figure 7 showing the depletion layer (from semiconductor theory) each side of the junction where the doping is n^+p type. What can you infer from this about efficiency in terms of location for recombination and resultant injection luminescence? In other words where are the photons created that improves efficiency?

External Quantum Efficiency

Whatever about internal value the *external quantum efficiency* for a basic surface emitting LED is much lower than 100% because of the difficulty of extracting all the radiation. Only light within a cone defined by total internal reflection TIR and the critical angle C can escape, Figure 8. The high refractive index n_1 for III-V compounds mitigates against extraction due to a small C. By Snell's law

$$n_1 \sin i = n_2 \sin r \text{ and } r = 90 \text{ deg before TIR}$$

Hence $\quad C = \sin{-1}(n_2/n_1) = n_2/n_1 \quad$ for small angles

Since n_2 is for air and n_1 is large the C value is small.

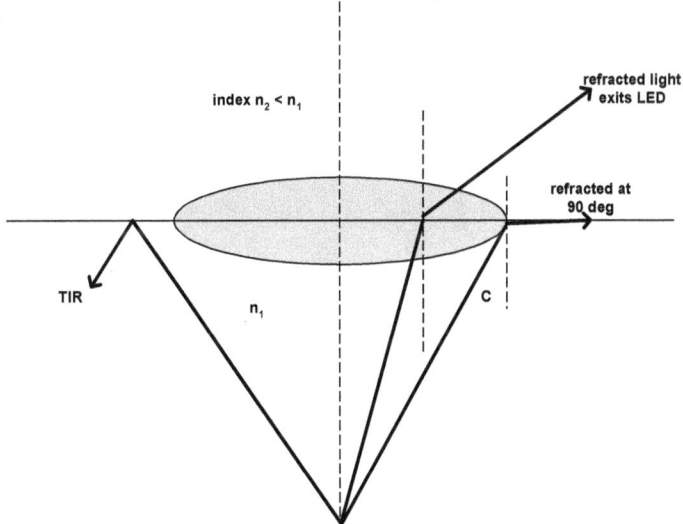

Figure 8. A cone of light defined by C escapes from the LED through the surface window; the rest suffers TIR.

Furthermore the fraction F that escapes is determined by the lit disc area relative to the surface area of a sphere filled uniformly with radiation since the emission is spontaneous and random in direction, Figure 9.

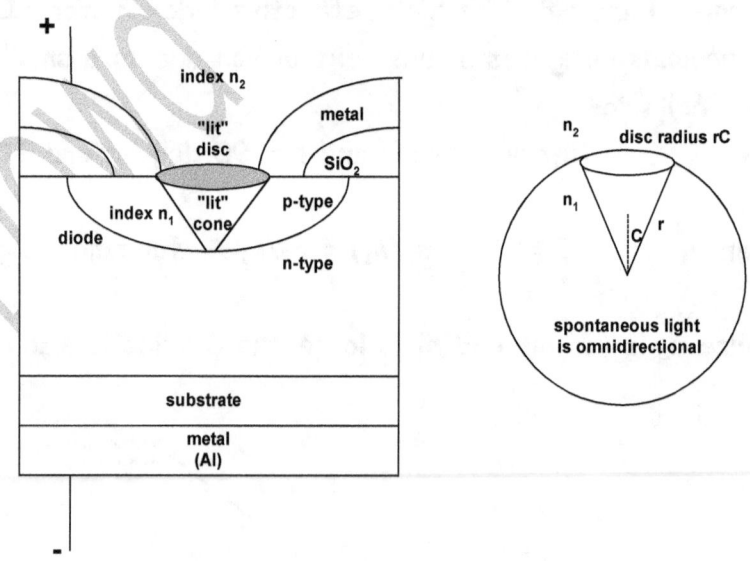

Figure 9. Cone of light that escapes through a lit disc of radius rC is determined by the critical angle C for GaAs.

Sphere surface $A_s = 4\pi R^2$ and lit disc area A_d is $\pi(rC)^2$ so that

$$A_d/A_s = C^2/4 = \tfrac{1}{4}(n_2/n_1)^2$$

Also we apply Fresnel's equation for transmission fraction at a boundary near normal incidence:

$$T = [1-(n_1-n_2)^2/(n_1+n_2)^2]$$

The total fraction extracted becomes:

$$F = \tfrac{1}{4}(n_2/n_1)^2[1-(n_1-n_2)^2/(n_1+n_2)^2]$$

This value is the external quantum efficiency for the surface emitter and is quite small, only a percent or two, as the following exercise shows.

You Do...
Ex 4. Surface emitter efficiency.

Show the critical angle for the GaAs surface emitter with refractive index 3.6 is small and the resultant external quantum efficiency is order 1%.

Answer Ex 4 at end handbook.

○

There is an immediate improvement in the external quantum efficiency if the upper p-layer surface were curved as a sphere to ensure rays strike at near 90 deg, Figure 10a. This classical optics trick is an expensive solution as it would require etching each chip to a dome shape. The alternative is shown in Figure 10b where a dome cap is added to the surface by encapsulating the chip in transparent plastic. Now the benefit of the spherical geometry is achieved more cheaply and additionally the refractive index of plastic around 1.5 helps greatly by altering the critical angle relative to an air interface. In the *Burrus-type LED* a pit is etched at the exit window to further improve extraction of light.

Exercise: Calculate the critical angles C for GaAs to plastic and for plastic to air.

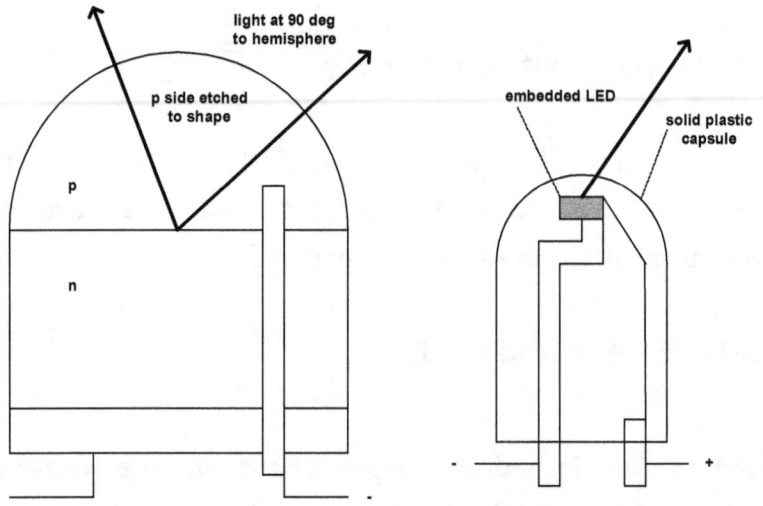

Figure 10. Hemisphere-shaped surface improves light extraction.

Edge Emitting LED, ELED

The deficiencies of the surface emitting LED are clear from the discussion so far. The *irradiance* is light power surface density launched into the forward direction, units $Wcm^{-2}sr^{-1}$. It is determined by injection current density, internal quantum efficiency, thickness of the recombination region and internal re-absorption since the GaAs energy gap also suits photon absorption. This combination of factors points to the edge-emitter design, Figure 11, as a great improvement, in fact x5 times better than a Burrus-type surface emitter. The modulation speed for the ELED is also superior. The irradiance for a Burrus-type surface emitter is about 200 $Wcm^{-2}sr^{-1}$ while for the ELED it is ~1000 $Wcm^{-2}sr^{-1}$ where the units refer to light density on unit area and with divergence contained within a solid angle of unit steradian. This specification is important for *coupling* the radiation into optical fibres as the more "contained" the light is the better and the system loss budget will be improved. It is also important for focussing light to a fine spot on a CD-ROM disc for *read/write* operation in the storage process.

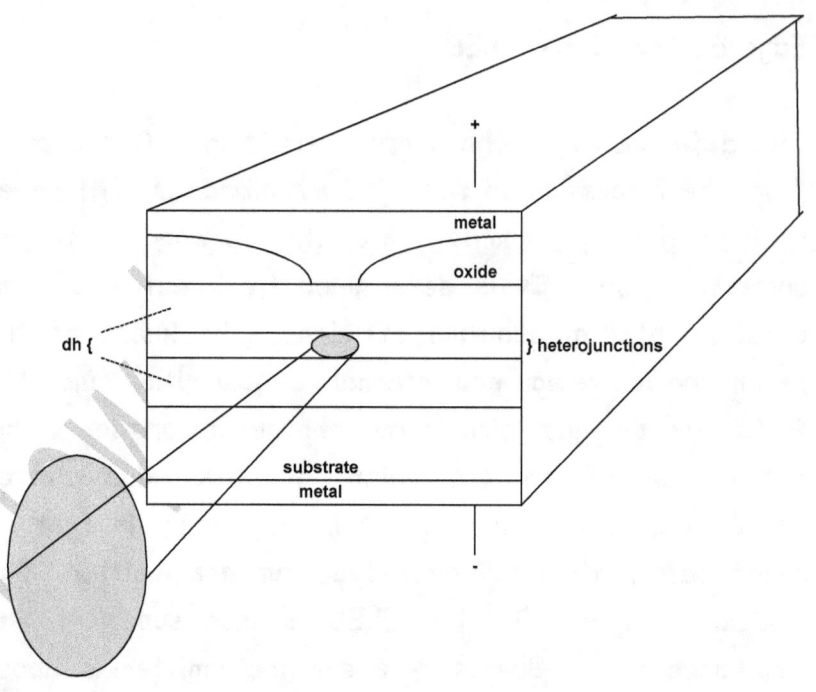

Figure 11. ELED comprised of ternary material AlGaAs with stripe contact and double heterojunction dh.

The AlGaAs ELED structure shown in Figure 11 comprises a substrate carrying a dh or double heterojunction wherein the pn diode processes take place and the resultant light is guided transversely down to the chip edge somewhat like in an optical fibre. This is called an in-plane design, referring to the junction plane. A very bright lit up window results at each end and the exiting light carries forward with it the "containment" features imposed by the semiconductor guide, namely improved irradiance. Above the dh layers there is an insulator oxide with a stripe-shaped gap and a metal layer contact on top that forces injected current into a descending sheet. That in turn provides high current density within the active region. The light arriving at the exit facets of the chip is now confined to a bright spot plotted as in Figure 12.

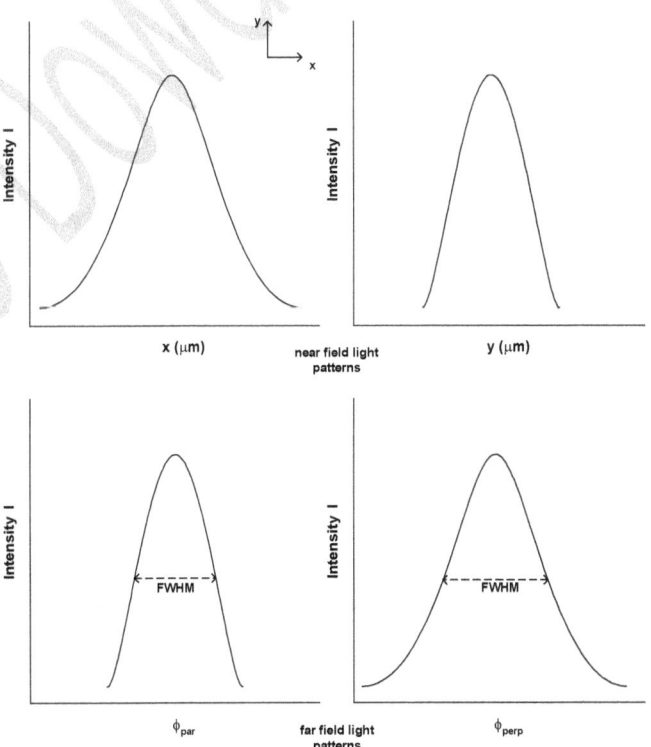

Figure 12. Near field (a) and angular far field (b) light patterns for a stripe contact dh ELED.

The *near field* pattern is one plotted as light intensity in Wcm^{-2} against distance in μm across the chip facet from one side to the other. The *far field* plots intensity against angle in degrees from the chip axis for directions parallel ϕ_{par} and perpendicular ϕ_{perp} to the junction. The latter is important for focussing onto the axis of an optical fibre. As the radiation now derives from a slit shaped lit window it diffracts as light would through a narrow slit, well known for classical optics. So the pattern at 90 deg to the plane of the slit, and equally the dh layers, spreads out more than that in the orthogonal direction as shown in Figure 12. The intensity falls off with angle to half its central value at angle 15 deg parallel to the junction but 25 deg perpendicular to the junction. Alternatively we say that the *full-width at half maximum* FWHM is 30 deg parallel and 50 deg perpendicular to the junction.

As for the near field we can infer that the lit spot at the exit facet will be oval, wider in the sideways than vertical direction. Typically this oval is only 1 μm vertical (y) by 10 μm sideways (x) and therefore a much improved tiny spot for focussing into fibre with 50 μm core or onto a storage disc. In the latter case a shorter wavelength gives even smaller *focussed spot-size* for high density data storage. We should therefore prefer

UV or visible blue light to IR for read/write operations and we must select different material band gap accordingly.

Double Heterojunction or Heterotructure

A common silicon diode consists of p-type and n-type doped Si forming a homojunction of the same material. For ELEDs we devise a *heterojunction* where the p and n sides consist of different alloys by varying the mole fraction x in $Al_xGa_{1-x}As$ for example or x and y in $In_xGa_{1-x}As_yP_{1-y}$ if a quaternary semiconductor is required. Here the mole fractions x and y determine the band gap E_g and resultant wavelength λ. The heterojunction structure allows us to tailor both the energy gap and refractive index variation across the layers thereby controlling electron and light confinement at the same time. A typical dh device is depicted in Figure 13 with five different alloys.

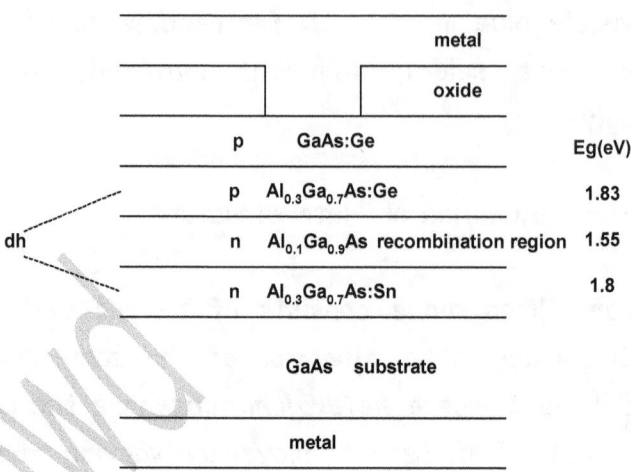

Figure 13. ELED dh structure using $Al_xGa_{1-x}As$.

For the active layer n-doped $Al_{0.1}Ga_{0.9}As$ is chosen with a band gap of 1.55 eV and recombination takes place here producing light at wavelength 800 nm. (Exercise: Verify this by calculation). This is guided by TIR until it reaches the end facets since the refractive index in $Al_{0.3}Ga_{0.7}As$ layers either side is lower. But since the energy gap 1.8 eV either side is greater than 1.55 eV the carriers are simultaneously confined to the lit region. Outside the dh structure there is the basic substrate layer below and a *Ge* doped p-type layer above for good ohmic contact on the p-side. A stripe at the metal-to-oxide contact ensures current *confinement* to a descending sheet. In all then there are three types of confinement, light, carriers and current, while re-absorption is also diminished by having a larger band gap than the photons on either side of the active layer. The double heterostructure greatly enhances the efficiency while creating a waveguide in the lit region.

Response Time for LED

The LED has a diffusion capacitance due to storage of carriers within a diffusion length or so of the junction. When applied voltage drops due to modulation by data these must diffuse away and then recombine to enable a new equilibrium to be established. For good frequency response the minority carrier lifetime τ must be small.

$$\tau = (Bp)^{-1}$$

Here p is majority concentration and B is the constant previously defined by r = Bnp which combined with r = n/τ gives the equation above. It is evident from this equation that τ may be reduced by heavy doping p thereby giving higher bandwidth. However near the solubility limit for acceptor impurities in GaAs non-radiative recombination centres are formed. With Ge in GaAs for example the external quantum efficiency drops above 10^{24} atoms per cubic metre. At that point using B from Table 3 above gives:

$$\tau = [(7 \times 10^{-16})10^{24}]^{-1} = 1.4 \times 10^{-9} \text{ s}$$

This result, 1.4 nanoseconds, suggests bandwidth around 700 MHz (the inverse) but in practice that is further reduced several times by stray effects from bond wires, packaging etc.

The alternative to high doping we will now show is heavy forward injection current density J with narrow active region thickness t. If Δp is injected carrier concentration and it greatly exceeds the equilibrium density p then

$$\tau = (B\Delta p)^{-1}$$

Now in equilibrium the number of recombinations each second in distance t is J/e (carriers per second per square μm) so that in unit volume the rate is J/te (carriers per second per cubic μm). But Δp recombine in 1 μm^3 in time τ so this rate is also $\Delta p/\tau$ and hence

$$J/te = \Delta p/\tau$$

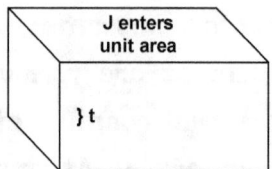

This produces $\quad \Delta p = J\tau/te$

Inserting this Δp into the above equation for τ gives:
$$\tau = (et/JB)^{1/2}$$

In order to reduce this τ value and achieve higher modulation speed we must reduce thickness t and raise injection current density J. Note however that since τ is now current dependent, or non-linear, this produces signal distortion as the modulation rises and falls.

Basic LED Drive Circuits

Two very simple schemes to derive modulated light from a LED are illustrated in Figure 14. For the first primitive circuit the LED specifications from the supplier data sheet are V_{ON} and I_{MAX} so a limiting resistor must be selected so that R_{LIM} maintains current within allowed range. When power supply voltage is set at V clearly R_{LIM} is $(V-V_{ON})/I_{MAX}$ and the switch may be manual or used to represent the modulation.

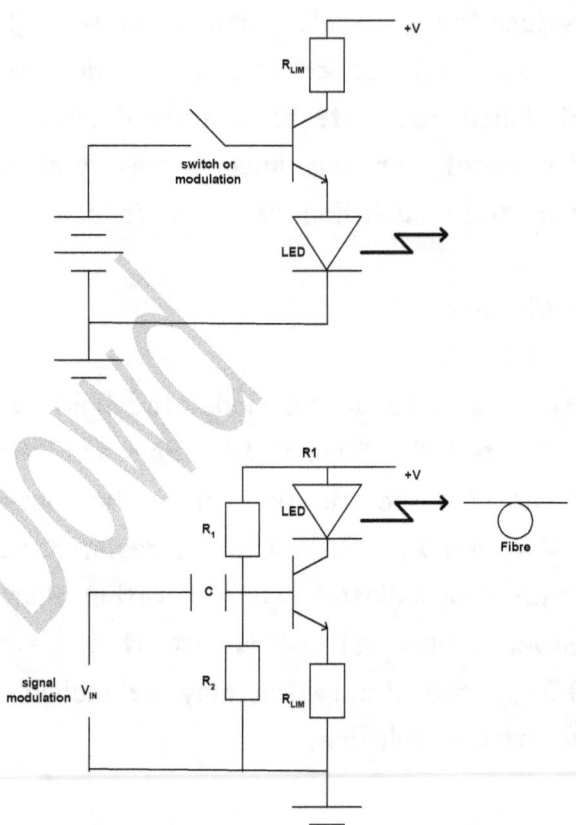

Figure 14. Primitive LED modulation circuits.

In the second scheme V_{IN} modulates diode current while R_1 and R_2 bias the transistor to half LED maximum. Both transistor and LED are in their linear regimes. The rear facet light of the LED could additionally face a photodiode that provides feedback on the level of optical power.

LED Modulation Bandwidth

The LED is normally deployed in a system comprising transmitter TX, receiver RX and optical fibre channel. Electrical injection I_{IN} supplies carriers so that photons are produced at a proportional rate and the optical power P_{OPT} is dependent on drive current I_{IN}. The electrical power however is proportional to I_{IN}^2 as shown in Figure 15.

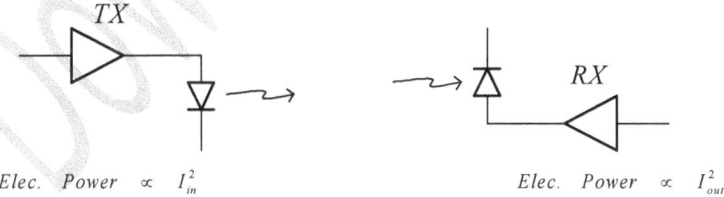

Figure 15. The optical fibre resides between two other sub-systems TX and RX.

The received light produces electrons by photon absorption so the electrical current I_{OUT} varies with the diminished P_{OPT} but the electrical power at RX depends as ever on I_{OUT}^2. The diminution of received P_{OPT} derives from the losses in the fibre and connectors.

Electrical regime dc-to-hf (high frequency) 3dB bandwidth bw is defined by that modulation frequency for which $I_{OUT} = I_{IN}/(2)^{\frac{1}{2}}$
Optical regime bw is defined by the modulation frequency for which RX optical power falls by $\frac{1}{2}$. This

implies IOUT has fallen also so it corresponds to an electrical power attenuation of 6 dB (factor 4).

If the response has Gaussian shape as expected then:

$$\text{Optical bw} = 2^{\frac{1}{2}} \text{ Electrical bw}$$

At the LED there is a time response to modulation of the drive current and this τ is combined with signal frequency ω or $2\pi f$ according to the Gaussian response function:

$$P(\omega)/P_{dc} = [1+(\omega\tau)^2]^{-1/2}$$

$P(\omega)$ is the mean modulated optical power with constant peak-to-peak current.

Pdc is optical power for the same dc drive e.g. 50 mA dc versus +/- 25 mA rf (radio frequency).

This means that optical power falls off relative to the output for a dc drive of the same amplitude whenever faster data modulation ω is deployed and equally with longer response time τ.

For edge emitters injected carrier density is the dominant factor for τ. Bimolecular recombination processes involving multiple carriers can reduce τ as do recombination at crystal defects:

$$N_2/\tau = AN + BN^2 + CN^3 = BN^2 \text{ approx.}$$
$$= \text{defects+radiative+Auger(phonon processes)}$$

B is given in Table 3 above and is $\sim 10^{-16}\,\text{m}^3\text{s}^{-1}$

Thin active-layer AlGaAs LEDs have electrical bw ~250 MHz but quaternaries have corresponding bw that is 2 to 3 times higher due to shorter τ where the recombination coefficient B is large and there are more non-radiative recombination centres.

The FDDI (Fibre data distributed interface) LAN standard (Local area network) at 1.3 µm window is above 200 Mbit/s so we must select quaternary LEDs for that.

Example: LED modulation response.
A LED has response time 5 ns and produces light output 300 µW for a given dc drive. If modulated at (a) 20 MHz and (b) 100 MHz what is the optical output? Calculate also the half power frequencies.

Answer. τ = 5 ns P_{OUT} = 300 µW   f = 20 MHz
$$P(20\text{ MHz}) = P_{dc}/[1+(2\pi 20 \times 10^6 \times 5 \times 10^{-9})]^{1/2}$$
$$= 254.2\ \mu W$$
Repeat using f = 100MHz: P(100 MHz) = 90.9 µW
For half power: $1/[1+(\omega\tau)^2]^{1/2} = \frac{1}{2}$ and $f = \omega/2\pi$
Hence $f_{optical}$ = 55.1 MHz $f_{electrical}$ = 55.1/(2)$^{1/2}$ = 39 MHz

3   Lasers

To operate a LED as a laser diode design we must visit some basics concerning *light amplification by stimulated emission of radiation* or l-a-s-e-r. Diode drive current populates the upper energy level with electrons and for laser action we require *population inversion* where the upper number N_2 exceeds that for the lower level N_1 as depicted in Figure 16.

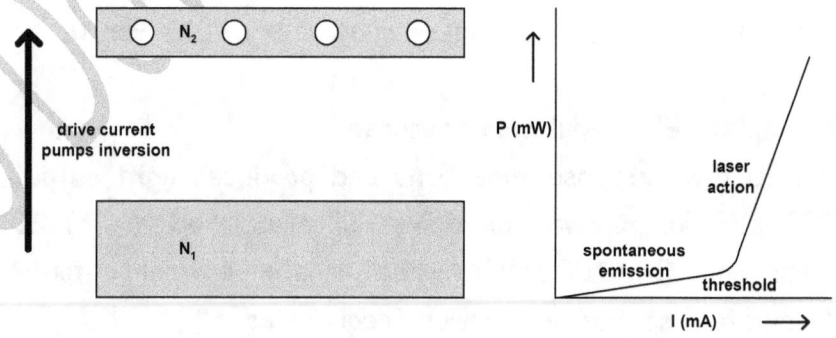

Figure 16. Population inversion in the semiconductor.

Next we expect some *stimulated emission* when a photon passes with energy corresponding to the gap E_g in the excited material. At this point we have a *semiconductor optical amplifier* or SOA because for each input photon we can get two out as recombination takes place, Figure 17.

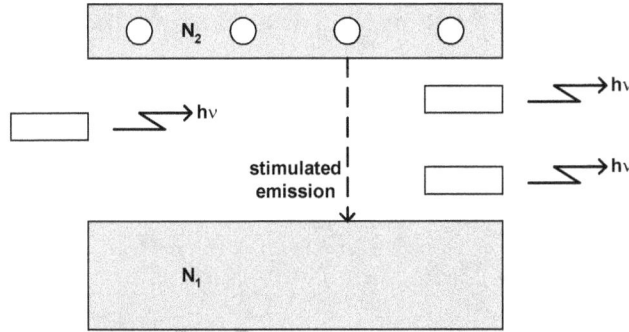

Figure 17. Stimulated emission produces a photon of equal frequency and energy to the injected photon.

To convert an amplifier to an oscillator it is known from electronics that feedback is required but in this case it should be *optical feedback*. A mirror at one or both ends will provide this. In our case the active material is a semiconductor so the partial mirror can be the polished or cleaved end facets of the chip. The partial reflectivity is found by Fresnel's equation and this two-facet mirror arrangement is called a FP or *Fabry-Perot cavity*. As the drive current is raised above a threshold the population inverts, stimulated exceeds spontaneous LED emission and the device starts to lase. This onset is accompanied by a rapid rise in optical power, Figure 16, a sharp narrowing of the optical spectrum around the energy gap frequency, Figure 18, and the appearance of coherent properties in the output light. *Cavity modes* are said to oscillate and these appear in the spectrum at higher resolution as lines for optical frequencies that are supported by the cavity length.

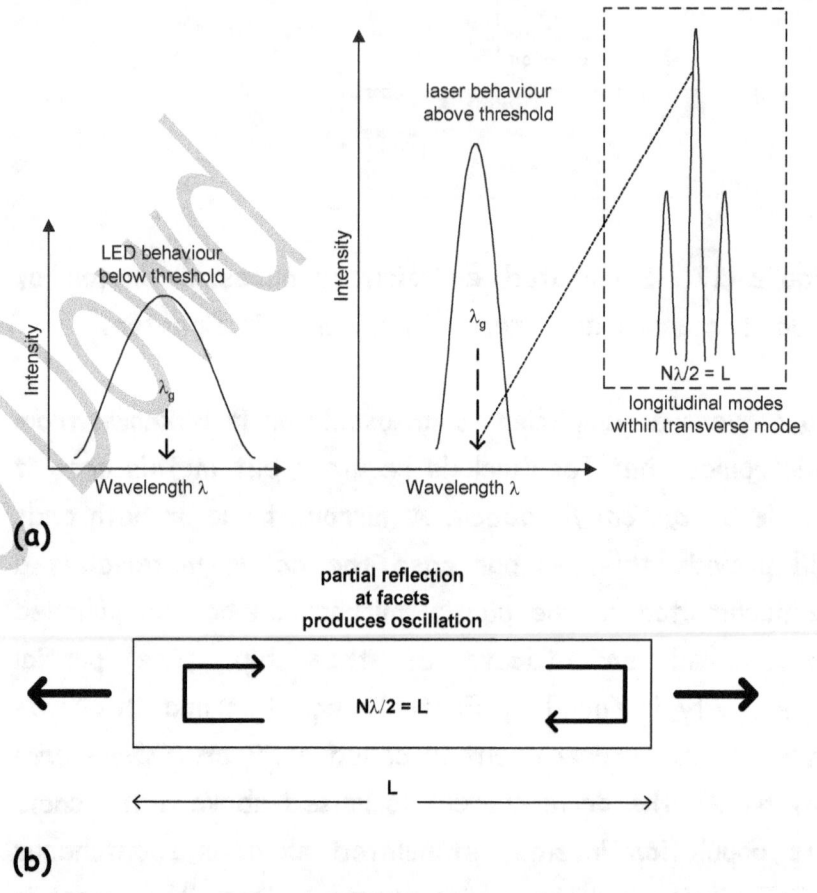

(a)

(b)

Figure 18. (a) Laser spectrum narrows sharply;(b) Chip with feedback facets that create an optical oscillator.

Einstein Equations

The three processes that may occur when light interacts with the material medium are photon absorption, spontaneous and stimulated emission. These were combined by Einstein using weighting coefficients A and B for each process and the populations for the lower and upper states N_1 and N_2.

The absorption may occur only when a photon field is present so it is stimulated and assigned B_{12}. Spontaneous emission is given A_{21} so that τ for that rate equals $1/A_{21}$.while stimulated emission is weighted B_{21}. Actual values depend on the properties of particular atoms and states according to molecular orbital theory.

In thermal equilibrium the populations of atoms were related by Boltzmann:
$$N_1/N_2 = \exp[(E_2-E_1)/kT] = \exp(\Delta E/kT)$$

Electromagnetic wave energy per unit volume per unit frequency span about frequency f is from Plank:
$$\rho(f) = 8\pi hf^3/c^3[\exp(hf/kT)-1]^{-1} \text{ black-body radiation law}$$

Stimulated absorption probability in interval δt is coefficient x density photons x time: $B_{21}\rho(f_{21})\delta t$
Stimulated emission probability in interval δt: $B_{21}\rho(f_{21})\delta t$
Spontaneous emission probability in interval δt: $A_{21}\delta t$

Rate = probability (per unit time) x number atoms available (per unit volume). At balance excitations and relaxations cancel:

$$A_{21}N_2 + B_{21}\rho(f_{21})N_2 = B_{12}\rho(f_{21})N_1$$

Hence $\rho(f_{21}) = A_{21}/B_{21}/[(B_{12}N_1/B_{21}N_2)-1]$

This by Boltzmann above gives:
$$\rho(f_{21}) = (A_{21}/B_{21}/[(B_{12}/B_{21})exp(\Delta E/kT)-1]$$

Compare this with Plank above and we must conclude:

$$B_{12}=B_{21}=B \text{ simply and also } A_{21}/B_{21}=A/B=8\pi hf_{21}^3/c^3$$

We also see that the ratio spontaneous/stimulated is:

$$A_{21}/\rho(f)B_{21} = [exp(hf/kT)-1] = \sim exp(hf/kT)$$
At 10^{14} Hz in the IR this produces
$exp(6.6x10^{-20}/4.2x10^{-21}) = \sim exp(16)$ or 10^5.

We conclude spontaneous emission is enormously more likely than stimulated in the IR. Hence the historical difficulty of getting laser action.
Conversely at rf of 10^8 Hz we get $exp(16/10^6) = \sim 1$ so the ratio above is (1-1)=0 meaning stimulated emission is certain for radio waves. A simple antenna is sufficient.

Exercise: At what frequency will the competing processes be equiprobable?

Answer: Put ratio = $A_{21}/\rho(f)B_{21}$ = 1 to get exp(hf/kT)=2 and thereby hf/kT= ~1. Hence hf/kT= ~1 or hf=kT
Now f = kT/h = $1.4 \times 10^{-23} \times 300/6.6 \times 10^{-34}$ = ~100 GHz in the far-IR or well beyond fibre optic windows.

Laser Threshold

The dh design is also used for semiconductor lasers and the prior Figure 13 is a typical case except that there is feedback from the end facets which are now cleaved to give a mirror finish and the device is driven above the threshold current I_t whose value will be determined next. The lasing cavity is formed by the junction which is along the <1,0,0> planes of the wafer. The orthogonal planes <1,1,0> are cleaved to produce equal reflectivity given by Fresnel where n=3.6 and n_{air}=1:

$$R_1 = R_2 = (n-1)^2/(n+1)^2 = 0.32$$

This 32% is more than sufficient feedback for laser action. Above I_t stimulated emission exceeds spontaneous and the LED behaves as a laser, Figure 16. The axial path then dominates whereas the LED operation had been omnidirectional. This gives a highly *directional beam* at the output except that at the narrow exit "slit" severe diffraction takes place and collimation with a lens may be required. Above I_t also, there is *spectrum narrowing* over ten-fold while uniformity of phase provides a *coherent beam* that exhibits *polarised cavity modes*.

Threshold is the point at which amplification of preferred modes by stimulated emission begins to exceed combined losses by absorption, scattering and optical losses at the output facets. In electronics this is where the oscillator is said to have closed loop gain of unity.

We will simplify the chip, Figure 19, in order to put the above statement of threshold into mathematics.

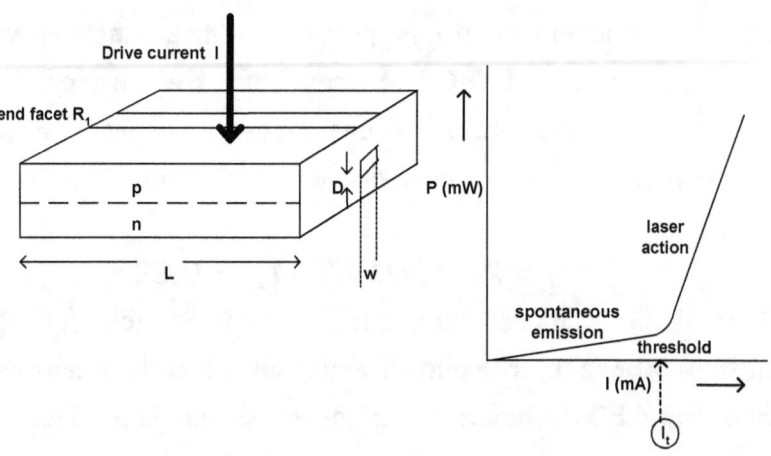

Figure 19. (a) Facets are separated by chip length L to form the lasing FP cavity. (b) Optical power versus drive current has a knee at threshold I_t.

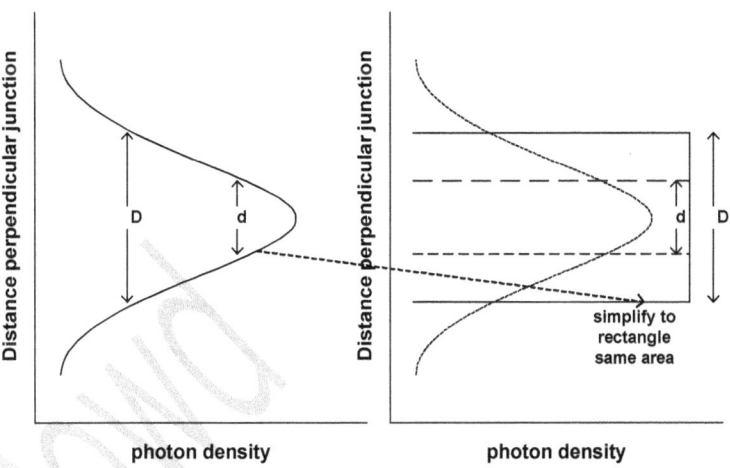

Figure 20. The near field with half power width D is simplified from bell to rectangle; d is active layer depth.

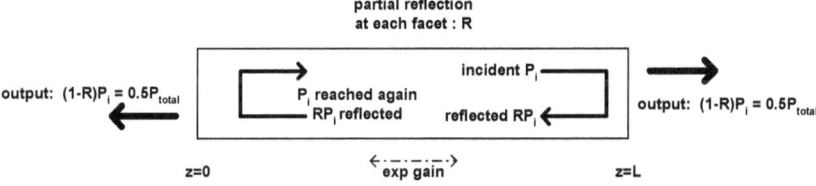

Figure 21. Light path up/down the cavity from z=0 to L. Total power is double the single facet output.

The light output at each chip end has a bell shaped near field intensity pattern that we simplify to a rectangle, Figure 20. Some light spreads out to extent D. The photons inside the active region of width d get amplified while those outside do not. All photons can suffer loss.

These exponential processes are summarised in Figure 21 to give power $P(z)$ at depth z inside the chip:

$$P(z) = RP_i \exp[(gd/D - \alpha)z]$$

Here the power P_i at a facet has portion R reflected at $z=0$ and that portion d/D is amplified at gain rate g while all light can suffer loss at rate α along the z-axis. For unity round trip gain RP_i recovers to P_i at $z=L$:

$$P_i = RP_i \exp[(g\Gamma - \alpha)L]$$

We use $\Gamma = d/D$ as confinement factor for the light restricted within the gain region. This produces:

$$\ln(1/R) = (g\Gamma - \alpha)L \text{ or } g\Gamma = \alpha + (1/L)\ln(1/R)$$

Use $g = \beta J$ where from physics $\beta = q_e \lambda_o^2 / 8\pi e n^2 d \Delta v$

to give $\qquad \beta J_t = \alpha + (1/L)\ln(1/R)$

$I_t = \text{Area} . J_t$ hence $I_t = (wL/\beta)[\alpha + (1/L)\ln(1/R)]$

J is injection current density (Am^{-2}) that provides and is proportional to gain g with constant β being device and material dependent.

q_e is internal quantum efficiency as before

λ_0 is vacuum wavelength at centre of laser gain spectrum

Δv is spontaneous emission spectrum width

n is gain medium refractive index

Exercise: Show by repeating the I_t derivation that when the facets have unequal reflectivity we use $(R_1R_2)^{1/2}$ in place of R in the above equation for threshold.

You Do...
Ex 5. Threshold current.
Find J_t and I_t for a GaAs laser of length 200 μm, injection stripe width 10 μm, uncoated facets. refractive index 3.2, confinement factor 0.8 where α = 10 cm^{-1} and β = 2.0x10^{-4} m/A

Answer Ex 5 at end handbook.
For Γ=1: J_t=3.35 kAcm^{-2} and I_t=67 mA
For Γ=0.8: J_t=3.35/0.8=4.19 kAcm^{-2} and I_t=83.7 mA

You Do...
Ex 6. External feedback.

R for the laser is effectively increased by 10% due to optical feedback from a nearby optical fibre facet. Re-calculate I_t for the above exercise.

Answer Ex 6 at end handbook.

○

You Do...
Ex 7. L-I or P-I plot.

Illustrate the effect on the P-I plot for this laser with and without the fibre feedback.
Answer Ex 7 at end handbook.

○

Laser Temperature Effects
At higher temperature T the threshold rises as shown in Figure 22 and the result is we can get "clipping" of the optical intensity modulation for the same bias current I_b. A higher bias added to the signal modulation is required to prevent this. A better alternative is to stabilise temperature with a Peltier-effect cooler. This is

commonplace in laser transmitters with temperature control within +/- 0.01 K. The Peltier-effect uses a thermo-couple type of device working in reverse; current across the bi-metal junction cools it.

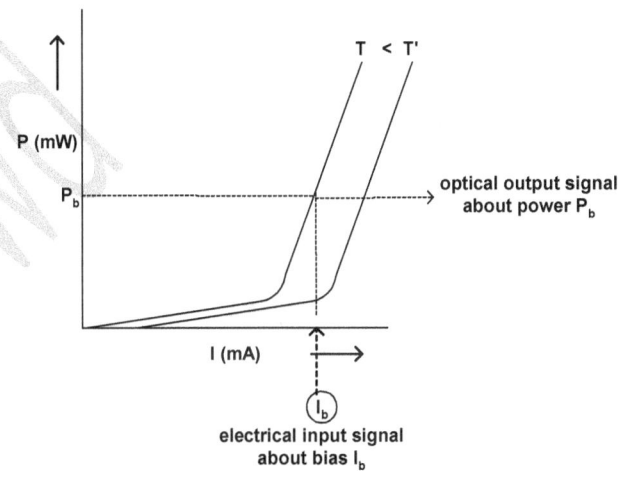

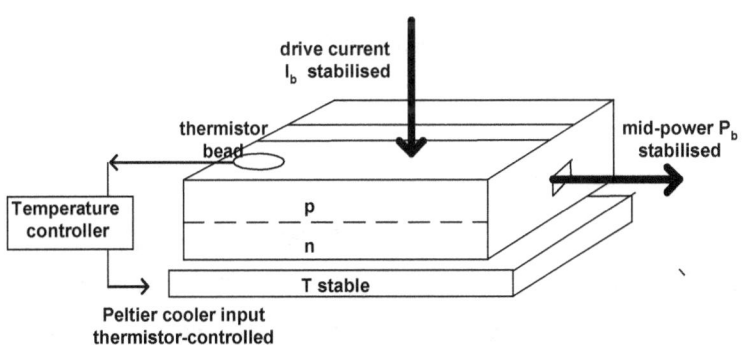

Figure 22. Upper: The P-I (or P-J) curve moves to higher current with increasing temperature T(K). Lower: Peltier-cooler stabilised laser chip.

$J_t(T)$ is influenced by variation in efficiency q_e, electrical and optical confinements and other factors. It is usual to fit an exponential to the combination as is expected of all thermal effects:

$$J_t(T) = J_0 exp(T/T_0)$$

Here J_0 and T_0 are empirically determined coefficients. Hence $(1/J_t)dJ_t/dT = 1/T_0$

We conclude that for low variation with temperature a *high T_0 is desirable*. Typically it is 150 K for ternary dh lasers but only 70 K for quaternary material due to q_e falling off faster with rising temperature.

A further effect is the wavelength variation with temperature $d\lambda/dT$, about 0.3 nm/K so that temperature regulation using a thermistor and Peltier cooler combination is essential for stable spectrum.

Laser Rate Equations

We now consider the two-level case applicable to semiconductor lasers where the dynamics are governed by coupled rate equations, one for carrier inversion density n and one for photon density s in the gain region. (Note: this is *not* the n also used for refractive index).

$$dn/dt = I_b/eV - Bns - n/\tau$$
$$ds/dt = Bns + F(n/\tau) - s/t_p$$

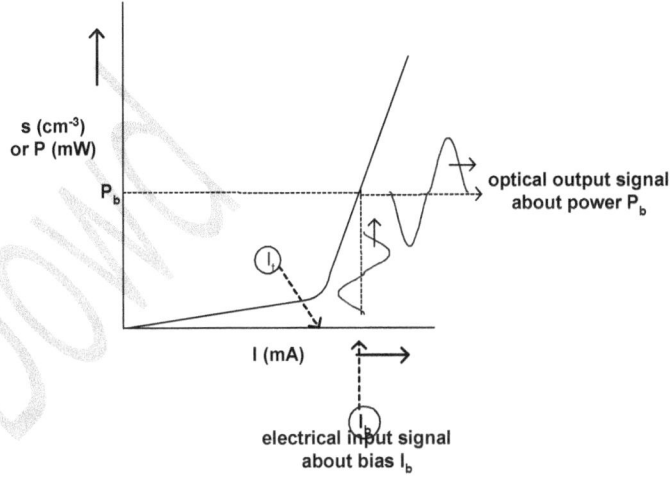

Figure 23. Plot of photon density s (cm^{-3}) versus injection current I (mA). This s mimics optical power P_0.

The first equation is the variation with time of injected carrier density due to bias current I_b in volume V where stimulated emission Bns and spontaneous recombination n/τ are meantime depleting the population. The *carrier lifetime* before spontaneous recombination is τ so the rate for this process is n/τ. The second equation gives variation with time of photon density where light is created by stimulated emission Bns and by fraction F of omnidirectional spontaneous emission $F(n/\tau)$ launched into the coherent direction. Simultaneously light s/t_p is lost at the end facets at a rate determined by t_p which is the average *photon lifetime* within the cavity before exiting. The common term that couples these differential equations is stimulated emission Bns as it is dependent on both variables n and s. The "pump" term is I_b/eV being the injection into volume V of carriers each of charge e. The fraction F is small, order 1%, so we can neglect it to see what is happening to n(t) and s(t).

Under the continuous wave output or "cw condition" ds/dt is zero giving:

$$n_t = 1/Bt_p \qquad \text{threshold carrier density}$$

Under the "threshold condition" and above it the carriers are at equilibrium density n_t so dn/dt is zero giving from the second differential equation along with known n_t:

$$s_b = I_b t_p/eV - 1/B\tau = (W - n_t/\tau)t_p$$

Here s_b represents coherent light output at the bias point I_b in the plot. W or I_b/eV is injection carrier density rate at the selected bias while n_t/τ is spontaneous recombination. The last equation subtracts spontaneous "waste" from the total and converts to coherent light output.

Example: Work out how to convert from s the photon density in photons per cm^3 to optical output power P_0 in watts. Hint: use Plank's equation and think of the "lit" inside volume V moving out into air at light group velocity c_g. We may conclude P_0 and s are therefore interchangeable.

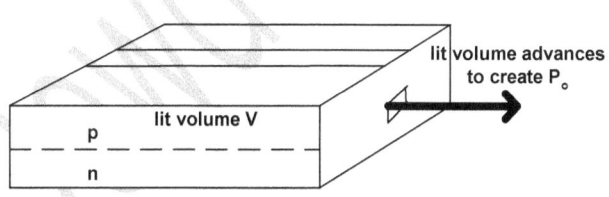

Answer: P_0 (watt) = $s(hc/\lambda)Vc_g$

Observe that for the plot of the last equation, s versus I_b, shown in Figure 23 and 24, we expect slope m in y = mx + C to be t_p/eV. Therefore we can *evaluate photon lifetime t_p* by measuring the slope of an experimental power versus current graph. It is typically some picoseconds while τ is a few nanoseconds. The intercept C is $1/B\tau$ so that measurement provides τ. For these P-I experiments (sometimes called L-I with L for light) we require to know device active volume V=wDL in Figure 19 and material coefficient B from Table 3.

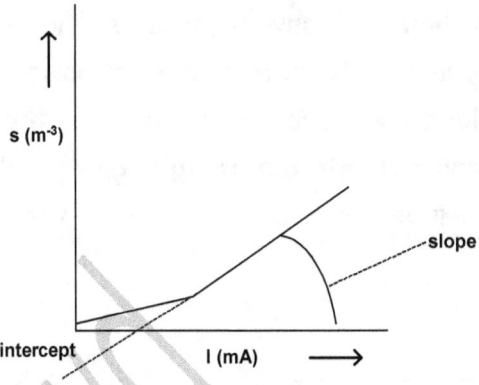

Figure 24. Slope and intercept for the s-I or P_0-I or L-I graph give photon and carrier lifetimes t_p and τ.

The lasing process can now be interpreted. At and above I_t or n_t a dynamic equilibrium is established that fixes n at value n_t so that dn/dt is zero (threshold condition). Additional pump then produces greater output, power rising with drive current, while carrier inversion remains fixed.

We may rewrite the s_b equation for a higher bias I_b'. Then consider modulating (I_{mod} = I_b-I_b') between I_b and I_b'. Now subtract these two equations to get modulated light s_{mod}:

$$s_{mod} = s_b-s_b' = I_{mod}t_p/eV$$

Hence the optical *intensity modulation index* or relative light intensity variation s_{mod}/s_b is:

$$m = I_{mod}\, t_p / eVs_b$$

Clearly a longer photon lifetime t_p or smaller active chip volume V improves data modulation capability.

You Do...
Ex 8. Photon density from power.

A typical communications laser emits 10 mW or 10dBm but a high power device is selected for 20 dBm as the link is atmospheric not fibre optic. It lases at 1.5 μm and the semiconductor alloy has average index 3.45 across the lasing spectrum. The device dimensions are stripe w=10 μm, lit depth D=1 μm, chip length L=100 μm. Calculate the photon density s.

Answer Ex 8 at end handbook.

o

4   Advanced Lasers

We have seen that temperature variations can affect the spectrum of the laser transmitter and that optical *frequency stability* is achieved using a Peltier cooler. More advanced lasers involve complex design that adds embedded optical control, for example using an in-chip diffraction grating. This operates somewhat like a conventional grating with many lines, not on glass however but etched into the semiconductor surface as corrugations. The "line" spacing in the grating section determines what wavelength is reflected back into the active section of the laser chip, Bragg's Law. Furthermore, when a current is supplied to the new section the charge density rises so the effective refractive index changes and with it the wavelength. This is then an electronically *tunable laser*. These complex features comprise advanced lasers for communications that can be deployed to exploit the full bandwidth of optical fibres shown in Figure 25. This shows the conventional C-band around 1550 nm wavelength centred on optical frequency 200 THz. There is also a dip near 1300 nm or 1.3 μm that is used for computer local area networks, LANs.

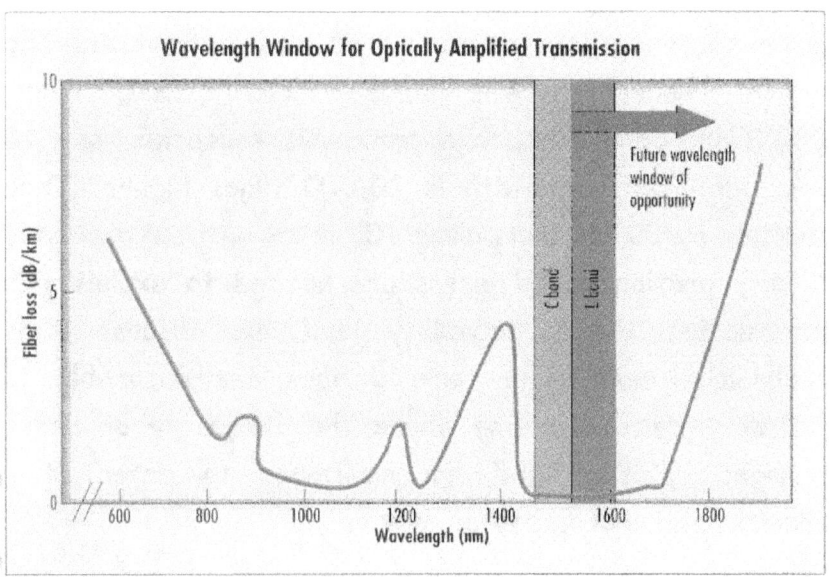

Figure 25. Transmission windows for SiO_2 or silicate optical fibre reside in the low attenuation parts of the loss spectrum. The conventional C-band at 1550 nm accommodates 50 or 100 dense optical frequency channels for DWDM.

The silicate glass made of extremely pure SiO_2 has tiny residual water contaminant that absorbs via the OH bond at 1400 nm in the near IR. This peak separates the short or S-band from the low loss *conventional or C-band* and beyond that lies the long or L-band. The C-band around 1550 nm wavelength is broad enough to accommodate 50 or 100 closely spaced optical channels. This practice is called *dense wavelength division multiplexing* DWDM. A frequency stable and tunable laser allows us to access all of this bandwidth. The optical frequency here is $\nu = c/\lambda$ or $3 \times 10^8 / 1.5 \times 10^{-6}$ giving 200 THz. Compare with microwave frequencies, say 10 GHz, and the bandwidth is 20,000 times higher. Then multiply by 50 channels gives 10^6 times greater capacity. Clearly grating based lasers are needed to exploit this information usage capability and we discuss that technology next. The same devices are applicable to *coherent optical sensing* where the frequency or phase changes of the light are monitored to detect tiny variations in a property of matter.

DFB Laser

This incorporates a Bragg grating embedded as corrugations in the layer above the active region of the laser chip and it replaces the facet reflector for feedback, Figure 26. Light leaks across the thin 0.1 μm layer to the grating where constructive interference by reflection occurs subject to the Bragg Law, integer times wavelength equals path difference. As the grating is spread along the length of the chip we have a DFB or *distributed feedback Bragg* construction.

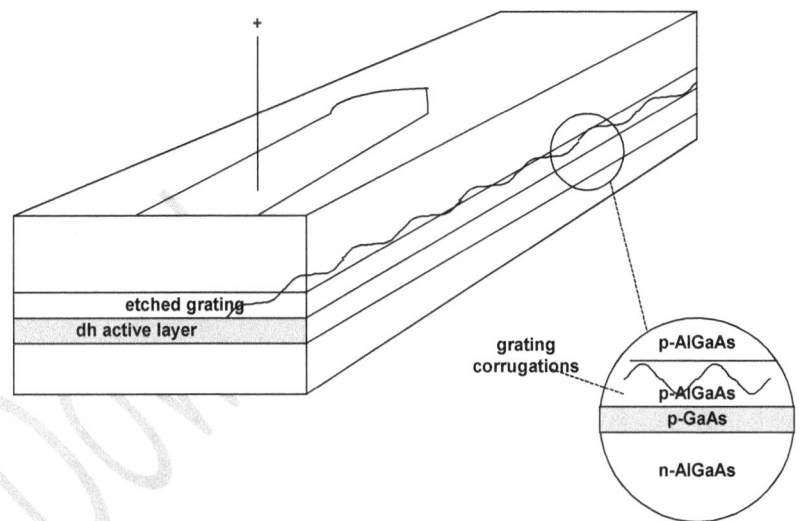

Figure 26. DFB laser structure.

The Bragg condition for reflection is:

$$N\lambda = 2n_eD$$

Here N is an integer, n_e is effective refractive index in the medium and D is corrugation period (separation of "lines"). In first order N=1, typical corrugation spacing is 0.25 μm and as index is about 3 we get λ=1.5 μm. The spectrum is now dependent on feedback according to this equation and frequency stability is improved ten-fold. The light FWHM spectrum spread Δλ is <1 nm and varies only 0.05 nm/K. Compare a Fabry Perot or FP device with two-facet reflection having $\Delta\lambda$ > 2 nm and temperature dependence 0.5 nm/K.

DBR Laser

An alternative design can have gratings at the end regions only, Figure 27. This is the *Distributed Bragg Reflector* variation.

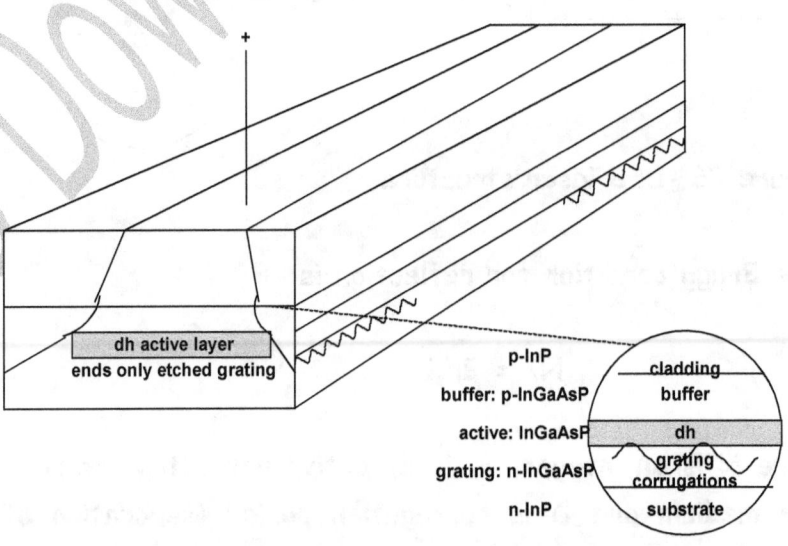

Figure 27. DBR laser.

High Power Devices

Normal lasing *spot-size* or near-field is typically 3 μm wide by 0.6 μm high while active layer thickness may be only 0.15 μm so that >50% of the light travels in the outer cladding. Such devices are limited to order 5 mW continuous-wave or cw power corresponding to 10 mW peak when intensity modulated with data (50% on). Higher power would damage the semiconductor end facet so it no longer acts as a mirror.

The TAL and LOC structures, Figure 27, are devised to overcome this limitation.

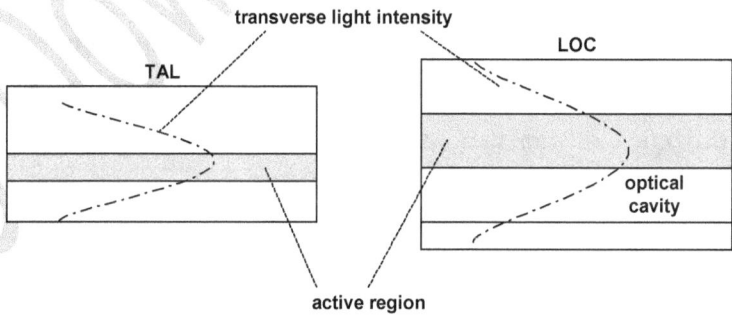

Figure 27. TAL and LOC structures for high power.

The thin active layer, only 0.05 μm, of a TAL laser causes the wavefront to diffract outwards thereby doubling the transverse spot to >1μm so 90% of light travels in the cladding. This provides up to 25 mW cw. The large optical cavity or LOC has a large guide of intermediate index adjacent to the active layer providing a spot ~1.5 μm with power ~40 mW cw. The large spot also reduces the far field beam which is determined by diffraction. A conventional AlGaAs spot of 3x0.6 μm gives an elliptical beam 12 deg × 45 deg. The large spot device produces ~7 deg × 25 deg.

Further techniques include (1) anti-reflection AR mirrors using multi-layer dielectric coatings to the facets and (2) non-absorbing mirrors of wide band-gap that can increase power ×5 times again to hundreds of mW.

Phased Array Laser, PAL.

It is possible to grow multiple lasers side by side on the same chip so that the beams combine to a powerful single output, a *phased array laser* Figure 29. The stripes at the output facet can couple via evanescent waves, analogous to a classical transverse diffraction grating where the phases lock together.

Light passing through each single slit produces a diffraction pattern, spread $2\lambda/s$, which is wider for a narrower stripe s. The multiple beams then interfere according to their separation period D with spread of the pattern λ/D. The product of these two patterns provides the output laser beam. As s is made closer in value to D the diffraction envelope encompasses only one or two interference lobes so we approach a single spot. For example a PAL with N=10, λ=850 nm and D=10 μm can have a FWHM for the central lobe about 0.5 deg and two lobes separated by ~5 deg. An array of 40 emitters with I_t=30 mA each or 1.2 A total can produce 2.5 W from a crystal only 0.15 mm^3 and internal lit volume of just 10^{-5} mm^3.

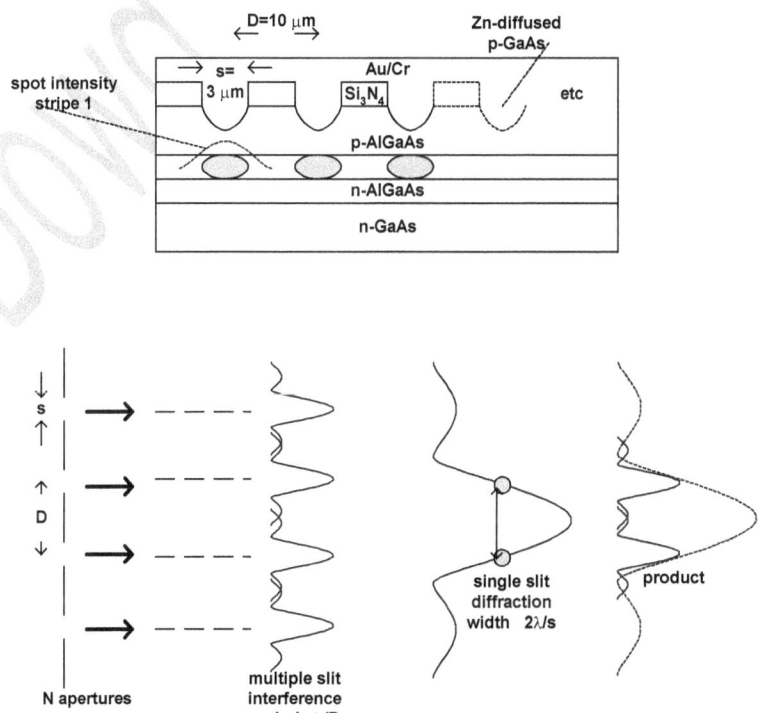

Figure 29. Multiple stripe phased array laser. Each stripe behaves like a slit and light from the N apertures diffract and then interfere.

Exercise: Calculate internal photon density s using the data for the last described laser.

Quantum Well Lasers

The potential well for an electron in the hydrogen atom is known from quantum mechanics and a similar arrangement, a *quantum well laser*, can be created artificially using a very thin layer for the active region within a semiconductor double heterostructure Figure 30.

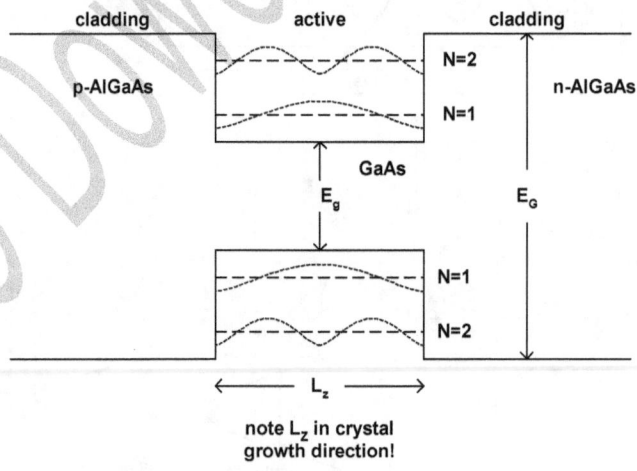

Figure 30. Electrical potential in a quantum well laser.

The width Lz of the thin active layer is chosen to be compatible with the electron wavelength λ_e as given by De Broghlie. Observe that the layers are drawn sideways in Figure 30. A conventional dh has gain region thickness ~0.2 μm but for the quantum nature to dominate it needs to be twenty time smaller, ~10 nm or 100 angstroms. Then the electron cloud is effectively a two dimensional gas along the hetero-boundary so there is very high electron mobility. (For that reason the same structure has migrated from photonics into the design of a high electron mobility transistor or HEMT for use in microwave communications). For the electron standing wave in the thin active layer as illustrated, an integer times half wavelength must equal the physical space so:

$$N \lambda_e / 2 = L_z$$

Use from physics electron momentum $p = h/\lambda_e$ and also motional energy $E_Q = p^2/2m^*$ where m^* is the effective electron mass.

Exercise: Show that these relations along with the equation above will produce separation ΔE_Q of the quantised energy levels. Hint: use N=1 and then N=2 and subtract.

Answer: Combine the three relations to get:
$$E_Q = N^2 h^2/8m^* L_z^2$$
Insert N=2 and N=1 and subtract for sub-level spacing:
$$\Delta E_Q = 3h^2/8m^* L_z^2$$

This gap relates to the closely-spaced stack of energy levels shown in Figure 31 for bulk and then QW material.

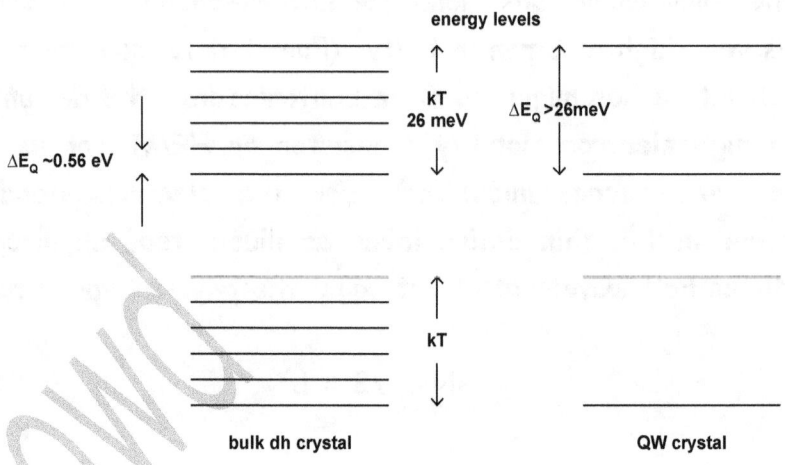

energy levels

ΔE_Q ~0.56 eV

kT
26 meV

ΔE_Q >26meV

kT

bulk dh crystal

QW crystal

Figure 31 Energy levels in bulk dh material, left, and artificial QW crystal, right.

When known constants and L_Z <20 nm are inserted into this result we find ΔE_Q which was 0.56 meV in a conventional or bulk dh laser now becomes >26 meV for the quantum well device, Figure 31. This result is crucial as it means the levels are now separated by more than the thermal energy kT=26 meV at room temperature. That value of kT was large enough to encompass the spacing of many close-packed levels in the *bulk material* but in artificial QW medium the levels have spread to the new ΔE_Q so only about two are occupied at room temperature. Population inversion is now achieved with ease and the threshold current I_t falls by decades. That in turn means less heating in the tiny chip, less spectral deterioration and reduced temperature dependence.

The smaller dimensions mean we require *nanotechnology* to grow the QW lasers and also diminished active volume suggests using several quantum well layers in a stack to get cumulatively enhanced optical power. This is a *multiple quantum well or MQW* laser. Outside the stack we grow dh layers to contain the light and carriers, the so called *separate confinement heterostructure SCH*. The nanotechnology may be achieved with advanced crystal growth such as metal-organic vapour phase epitaxy, MOVPE.

Vertical Cavity Surface Emitting Lasers, VCSELS

Although edge emitters are widespread in fibre optic communications there has been a return to surface emitting technology for CD players, laser printers etc, since the advent of QW capability. These VCSELs (pronounced "vicsels"), illustrated in Figure 32, have smaller spot size and narrow beam so they also can be more readily focussed into a fibre.

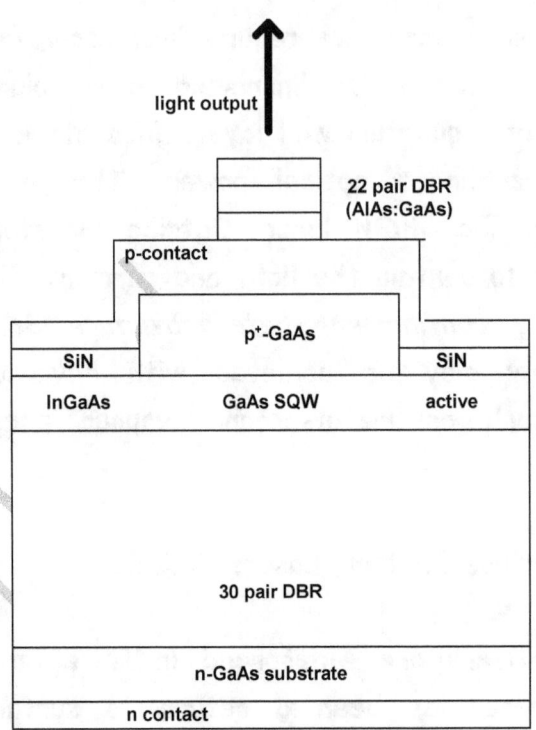

Figure 32. VCSEL with DBR mirrors.

The mirrors of the FP cavity are above and below the single quantum well, SQW, light-creating region. The mirrors can be made of multiple semiconductor layer-pairs of high and low index to form DBR reflectors. These pairs replace the corrugations of prior bulk DBRs. This VCSEL device is a triumph of nanotechnology and experiments have also shown it can be deployed at above 100 Gbit/s data rate when the spin of injected electrons is polarised. A great advantage of surface emitters is the individual lasers may be inspected for L-I plot before dicing into chips; wafer inspection is more economical than chip-by-chip for mass production.

Nanotechnology, Artificial Silicon, Optical Computers

The quantum well already described can be extended to two dimensions creating a quantum wire, or even to 3D resulting in a quantum box within the crystal called a *quantum dot*. The threshold current is thereby reduced significantly with even lower power consumption. But there are further benefits to these artificial crystalline materials. An electron may orbit a hole similar to a hydrogen atom with a nuclear proton. This carrier pair in a semiconductor is an *exciton* and in such a state the electron is localised to a position x with less room Δx to manoeuvre. By the Heisenburg Uncertainty Principle $\Delta x \Delta p$ relates to the Plank constant h so smaller Δx means greater Δp. That implies, when considered in the context of Figure 4 for indirect band gap silicon, that a significant momentum shift Δp is more probable facilitating a luminescent recombination even in Si. Now consider a Si crystal grown with a matrix of quantum dots. As electrons will be trapped in the quantum well boxes they are localised so the artificial crystal facilitates radiative recombination. The dream of using the most common element in the Earth's crust, Si, for lasers can become a reality.

These materials also exhibit a refractive index n that varies with light intensity I so that higher order n(I) terms must be added to our equations and a vista opens for *non-linear optics*. This provides for parallel photonic processing using optical wavefronts and for ultrafast optical computers.

Tunable Laser

The DBR device depicted in Figure 26 has an oscillating wavelength governed by Bragg's Law:

$$N\lambda = 2n_eD = f(n_e, D) \qquad \text{Bragg's Law}$$

This indicates we can control the light by both etching the corrugations in the end grating reflector to the correct period D and by effective refractive index n_e. This latter is determined by the alloy in the reflector layer but there is a further dependency; an injected current can raise the carrier density, changing n_e, and that will cause the feedback wavelength to shorten. This means we have an *electronically tunable laser* with a gain current I_A to the active layer section for power setting and a tuning current I_G to the DBR grating section for frequency control. A third section for optical-phase control can also help, and indeed a fourth section as DBR reflector at the other chip end. This construction was devised by the team led by Coldren at the University of California; these controls are the phase current I_P and second grating I_{G2}. The combination $(I_A, I_{G1}, I_{G2}, I_P)$ was found to give tunability across a full quarter of the C-band so that only four lasers in the transmitter TX could cover the entire ITU C-band comb of 100 optical channels. This opens the way to agile optical frequency allocation and consequently *"bandwidth on demand"* for customers of Internet Service Providers. The detailed operation of this laser is left as a Library Exercise (see O'Dowd, IEEE Jnl Selected Topics in Quantum Electronics, Special Issue March 2001).

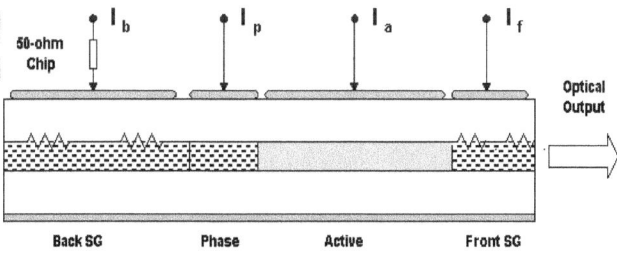

Example of a tunable laser with back and front gratings and four tuning currents. (Coldren et al, UnCalSB).

This type of tunability also permits passive *wavelength routing* of data through a network since the tuned IR-colour can be the packet's address and optical filters, like gratings, at each node can direct it do destination.

5   Photodetectors

Silicon in its common or natural form was forsaken in our search for luminescence but it is certainly useful in receivers for the detection of light as long as the wavelength is shorter than 1100 nm corresponding to the energy gap. Si is transparent to and absorbent of higher frequencies. This restricts it to UV, visible and very near IR. At longer wavelengths, especially C-band, we require III-V compounds in photodiodes.

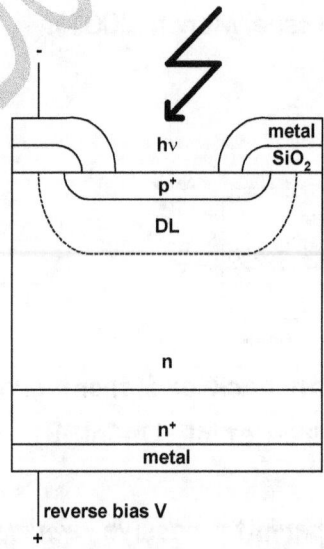

Figure 33. Silicon pn diode photodetector.

The simplest structure is a p⁺n diode having a heavily doped p⁺ side, Figure 33. With a p⁺ layer in top of the n region the depletion layer (DL) extends well into the n side as shown but is compacted on the p side. Below there is an n⁺ substrate for good ohmic contact with the metal. A window above in the SiO_2 silicate insulator allows light photons, $h\nu$, to enter. After photon absorption and under *reverse bias* the electron hole pairs are created within a diffusion length of DL and swept out to provide a photocurrent. The absorption coefficient is high at shorter wavelengths. At longer wavelength λ a wider DL is needed to ensure absorption implying higher reverse bias V but eventually that will exceed the breakdown voltage. The solution is a p-i-n structure or PIN diode with a wide intrinsic i-layer, Figure 34.

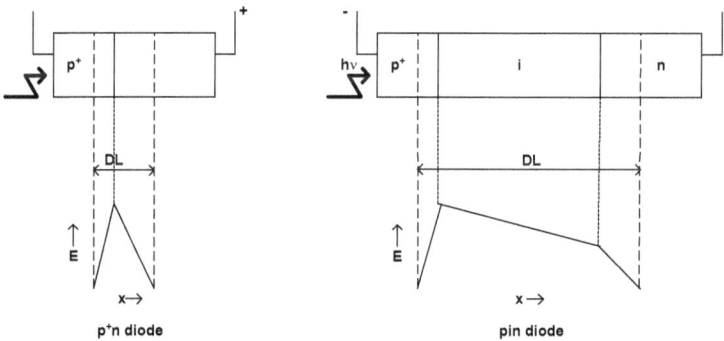

Figure 34. PIN diode with intrinsic Si central region. The internal E-field extends through the i-layer.

The idea here is that intrinsic silicon being un-doped has relatively high resistivity so the internal E-field falls slowly as depicted and there is effectively a wide absorption volume created as photons penetrate this. Only a few volts are required for DL to penetrate right through so that common logic processor levels are sufficient, e.g. 5 V for TTL. This design is useful at the longer wavelengths where the absorption coefficient is small. An additional benefit is that the capacitance associated with the junction, $C=\varepsilon A/d$, is diminished at longer d. That provides for shorter lifetime, $\tau=RC$, *faster response* and therefore wider bandwidth. The operation whereby a photon creates an electron-hole pair that is swept out by the internal E-field is depicted in the energy level diagram, Figure 35.

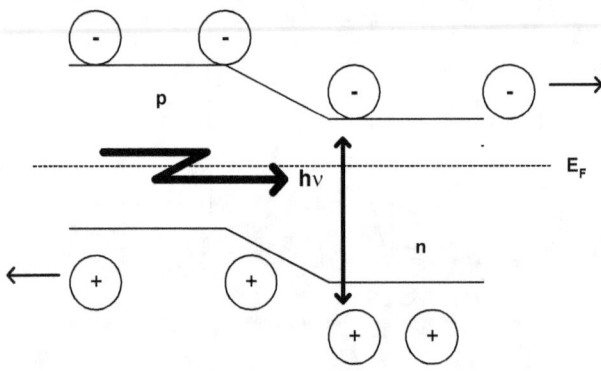

Figure 35. Energy levels in a photodiode and photon absorption creating electron-hole pairs that are swept out by the internal E-field.

On open circuit a potential difference appears at the output terminals. This is called *photovoltaic mode PV*.

When an external diode reverse bias V_d is used a photocurrent is available internally i_λ and portion i_{ext} is available at the terminals, Figure 36. This is called *photoconductive mode PC* and is most common in communication receivers, RX.

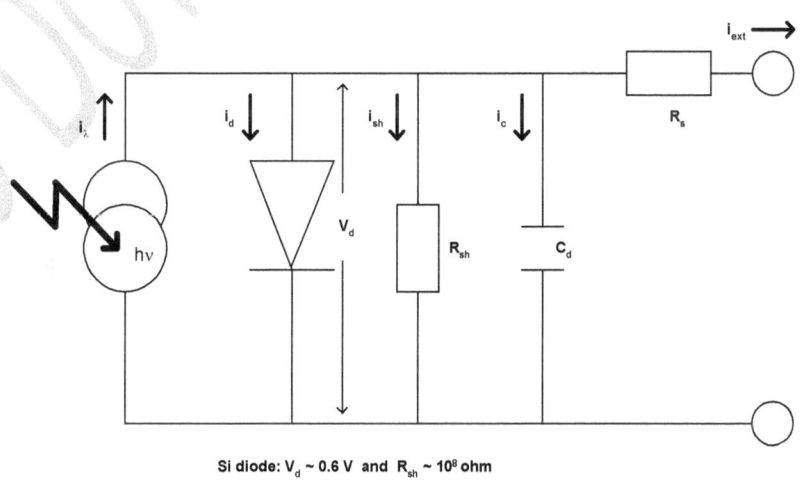

Si diode: $V_d \sim 0.6$ V and $R_{sh} \sim 10^8$ ohm

Figure 36. Equivalent circuit for photodetector.

If all incident radiation is absorbed within the photocell then each photon $h\nu$ creates an electron in the photocurrent i_λ:

$$i_\lambda = q_d(I_{opt}A)/h\nu)e$$

Here q_d is the diode efficiency, A is the window area, e electronic charge and I_{opt} is the optical energy density arrival rate called *irradiance* with units $Jm^{-2}s^{-1}$ or Wm^{-2}. For example sunlight irradiance at European latitudes is $\sim 1\ kWm^{-2}$.

Sensitivity

The equivalent circuit in Figure 36 has the photocurrent generator at left delivering current i_λ that breaks up into diode current i_d, shunt resistor current i_{sh} and diode capacitance current i_C. Only remainder i_{ext} flows through the series resistance R_s and is available externally for amplification and processing. In first evaluation of the photodiode *sensitivity* neglect the model's current i_C through the capacitance, valid at low modulation frequencies. Then apply Ohm's law, Kirchoff's law and the common diode equation as follows:

$$i_\lambda = i_d + i_{sh} + i_{ext}$$
$$V_{ext} = V_d - R_s i_{ext}$$
$$i_d = i_0[(\exp(eV_d/kT) - 1]$$
$$V_d = R_{sh} i_{sh}$$

In PV mode i_{ext} is very small and can be neglected by comparison giving:

$$i_\lambda = i_d + i_{sh} \text{ and } V_{ext} = V_d$$

Hence $\qquad i_\lambda = i_0[\exp(eV_d/kT)-1]+V_d/R_{sh}$

$$\exp(eV_d/kT) = 1+i_\lambda/i_0-V_d/R_{sh}i_0$$

For a silicon diode V_d=0.6 V and R_{sh}=10^8 ohm . Typically reverse leakage current in the diode equation is only ~10 nA so the last equation simplifies to:

$$\exp(eV_d/kT) = i_\lambda/i_0 \text{ since } i_\lambda >> i_0$$

Natural log of each side produces:
$$V_d = (kT/e)\ln(i_\lambda/i_0)$$

Finally using V_{ext}=V_d for parallel branches along with the original photocurrent i_λ equation gives:
$$V_{ext} = (kT/e)\ln(q_dI_{opt}Ae\lambda/hci_0) \quad \text{PV MODE}$$

This tells us the external voltage in PV mode is a log function of incident light irradiance. That mode is used in solar cells and the power resource for that at 50 deg latitude is $I_{opt} \sim 1$ kW/m^2.

PC mode
Typically ~10 V reverse bias is applied to operate in photoconductive mode for telecommunications links so the diode is well into the reverse saturated characteristic and i_d=i_0 giving from the first equation for i_λ:

$$i_\lambda = (i_0 + i_{sh}) + i_{ext}$$

The bracketed current is termed the "dark" current as it is present even when no light or data signal is present. It represents noise in the receiver. Again for a Si diode $i_{sh} = V_d/R_{sh} = 10/10^8 = 100$ nA while i_0 is ~10 nA and i_λ is typically ~1 μA. This order of magnitude comparison of terms allows us to simplify:

$$i_{ext} = i_\lambda = q_d I_{opt} A e \lambda / hc \qquad \text{PC MODE}$$

In photoconductive mode the external current available for processing is directly proportional to the incident light irradiance I_{opt}. This means an intensity modulated IM laser will provide a linear response at the RX. We see that response is also proportional to wavelength λ.

The PC mode has a *linear response* to light signal and is also faster, has better stability and greater dynamic range than PV mode. The *dark current* however $(i_0 + i_{sh})$ gives rise to *shot noise* which limits ultimate receiver sensitivity.

Responsivity of Si

The *responsivity* is depicted for silicon in Figure 37 showing no response to light beyond 1.1 μm as the band gap there is too large to absorb these smaller photons.

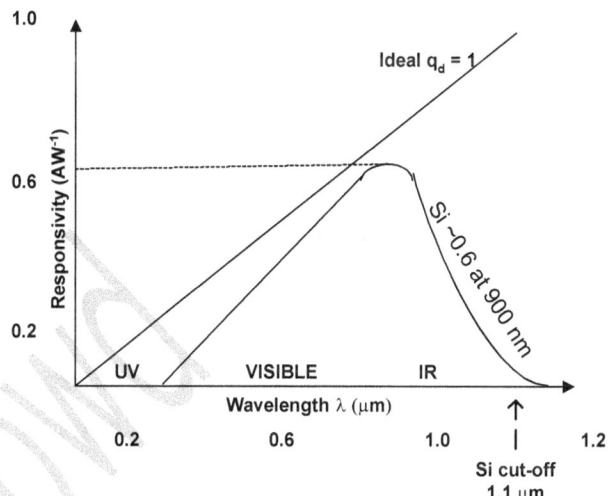

Figure 37. Responsivity spectrum for Si photodiode. Other materials have similar shape, different cut-off.

Photocurrent i_λ was defined using diode material efficiency q_d and optical irradiance I_{opt} in Wm^{-2} as follows:

$$i_\lambda = q_d(I_{opt}A/h\nu)$$

We define photodiode *responsivity* as $i_\lambda/I_{opt}A$ or photocurrent in amp produced per watt of received light.

$$\text{Responsivity} = i_\lambda/I_{opt}A = (q_d e/hc)\lambda \quad \text{units: } (AW^{-1})$$

The ideal efficiency is unity (or 100%) but in practice it can be ~0.6 at 900 nm when using Si. That means the ideal straight line plot shown in Figure 37 is not attained using Si but rather is about 60% lower stretching from the UV at 300 nm to 1100 nm in the near IR where the band gap cuts it off rather abruptly. The units of responsivity are often quoted in $mAmW^{-1}$ to reflect actual currents achieved in a fibre optic link and crucially we see it is proportional to wavelength... *the responsivity grows with* λ. The cut-off is at 1.1 μm for Si but 1.8 μm using germanium Ge and lies beyond 1.6 μm for selected III-V alloys so they can be deployed for C-band systems where their responsivity is closer to ideal.

Exercise: Responsivity dependence on wavelength.

Try to explain why the response should rise linearly with wavelength in the ideal photodiode case.

Hint: estimate number of photoelectrons created for each watt of received light starting from Plank's law.

The properties of common photodetector materials for different photonic applications are shown in Table 5. The last column is ratio of the ionisation coefficients $k=\alpha/\beta$ for electrons to holes and is useful for the avalanche effect that we consider next.

Table 5. Properties of photodetector materials.

Material	Bandgap (eV)	Cut-off λ		Ratio $\alpha/\beta = k$
Si	1.1	1.1 μm		10-100
Ge	0.67	1.8		0.5
GaAs	1.43	0.85		1-0.01
$GaIn_{86}As_{14}$	1.15	1.1		0.25
$GaIn_{47}As_{53}$	0.75	1.6	note!	5
InAs	0.33	3.8		This column
InGaAsP	1.34-0.78	0.92-1.6	!	electron/hole

Avalanche Photodiodes APDs

The avalanche photodiode provides internal gain of the received photocurrent and uses for this a structure similar to the PIN. High reverse bias produces avalanche multiplication in the I-region. Average multiplication factor <M> is ~10 to 100.

Automatic Gain Control AGC

Since the ionisation coefficient for electrons and holes, α and β in Table 5, are critically temperature dependent the reverse bias requires control for thermal drift. This will give more uniform avalanche gain. Also α and β are exponentially related to E-field so <M> is sensitive to voltage and that must be automatically controlled within

+/- 10 mV. The gain process is statistical with a spread of multiplications for pairs of electron-hole. The ratio k of α to β in Table 5 or its inverse 1/k is selected depending on which particle is dominant in the avalanche process for that material. The variation of gain values contributes noise to the photocurrent. This results in an excess noise factor f(M) empirically modelled as follows:

$$f(M) = M[1-(1-k)(M-1)^2/M^2]$$

We see that f(M) reduces with k and is worst when k=1. In the case of Si we use 1/k in the model as holes dominate the avalanche in that semiconductor. An alternative power-law fit to experimental behaviour is often used:

$$f(M) = <M>^x \text{ where } x=0.3 \text{ to } 0.5 \text{ for Si}$$

In practice we deploy an APD when the detector *shot noise* is well below the *circuit thermal noise*. Then optimum *signal-to-noise ratio SNR* occurs when the multiplication process, controlled by the voltage, brings shot noise up to the same level as thermal noise. Further increase in M will then deteriorate SNR. This will be shown next.

Receiver SNR

Consider a transmitter TX with optical power modulated sinusoidally at frequency ω about average power P_t with a modulation index m:

$$P(t) = P_t(1+m\sin\omega t)$$

Any more complex data signal, such as a square wave, can be broken into a Fourier sum of sine waves so this analysis is equally valid. The RX avalanche photocurrent, neglecting distortion due to fibre dispersion, will be:

$$I = I_0 \langle M \rangle (1+\sin\omega t)$$

Here $I_0 = q_e e P_r / h\nu$ is the average signal current before multiplication (*not* irradiance I_{opt}) and P_r is the received optical power after fibre loss. With a PIN diode use $\langle M \rangle = 1$ and $f(\langle M \rangle) = 1$ in what follows. Neglecting multiplied dark current the total average noise in the system from electronics coursework is:

$$\langle i^2 \rangle = \langle i^2 \rangle_c + 2eI_0\langle M \rangle^2 f(\langle M \rangle)B$$

Here $\langle i^2 \rangle_c$ is mean-squared non-multiplied circuit noise, the added term is shot noise associated with I_0, and B is effective noise bandwidth. The shot noise has $\langle M \rangle^2$ since I_0 is multiplied first by $\langle M \rangle$ and afterwards the noise on this is itself multiplied by $\langle M \rangle f(\langle M \rangle)$. Defining

SNR in terms of the ratio of mean squared signal to mean squared noise currents:

$$SNR_{APD} = \tfrac{1}{2}.m^2 <M>^2 I_0^2/[<i^2>_c + 2eI_0 <M>^2 f(<M>)B]$$
$$SNR_{PIN} = \tfrac{1}{2}.m^2 I_0^2/[<i^2>_c + 2eI_0 B]$$

This is because averaging the mean-squared sinwave produces $\tfrac{1}{2}$ at the numerator.
The amplifier circuit thermal noise is known from electronics coursework to be:
$$<i^2>_c = 4kTB.F/R_L$$

Here F is the noise factor of the input electronic pre-amplifier or low-noise amplifier LNA and R_L is amplifier input resistance.

Digital and Analogue Photodetection

For a given m, B and LNA, i.e. fixed $<i^2>_c$, the SNR equation above is a function of I_0. When I_0 is small so is its shot noise so the circuit noise dominates:
$$SNR_{PIN} = \tfrac{1}{2}.m^2 I_0^2/<i^2>_c \qquad \text{Circuit noise limit}$$
This implies low SNR and therefore a *digital system* where only two binary levels need to be discerned. Conversely when I_0 is large the shot noise dominates:
$$SNR_{PIN} = m^2 I_0/4eB \qquad \text{Shot noise limit}$$

This applies to an *analogue system* since large I_0 is required to discern multiple levels and faithfully reconstruct the signal shape.

As the latter expression is independent of circuit noise it applies to a perfect electronic construct (ideal LNA) so it represents the *FUNDAMENTAL QUANTUM LIMIT* for sensitivity of analogue systems or even a digital system with perfect LNA.

Optimum SNR for APD

Consider the APD expression above for SNR with initially $<M>=1$ where is behaving as a PIN. Now raise the reverse bias voltage so that SNR increases with $<M>^2$ for a while as we leave the circuit noise limit. Eventually however the growing shot noise equals circuit noise. Thereafter SNR falls off again with $[f(<M>)]^{-1}$.

Exercise: Alternatively, remembering $<M>$ depends on voltage, you may differentiate the SNR_{APD} expression with respect to $<M>$ and optimise by setting $d/d<M>$ at zero. The two noise expressions will then be equal. Use U/V derivative expression $(VdU-UdV)/V^2$

An alternative to the APD is a monolithic chip that matches a LNA to a PIN diode and because of miniaturisation exhibits overall excellent SNR. The inbuilt amplifier is often a field-effect transistor and this is a PIN-FET photo-receiver, Figure 38.

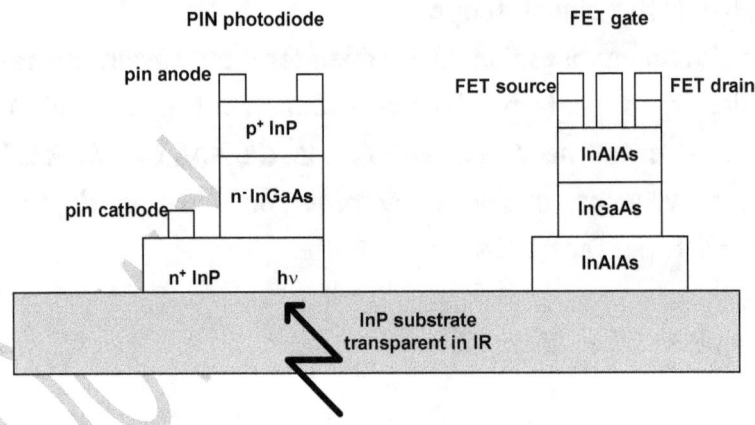

Figure 38. PIN-FET photodetector receiver.

Summary

We have learned to select photonic materials that match our requirements by band-gap engineering. Injection luminescence was deployed in various LED structures and then in semiconductor lasers. Advanced lasers were studied that cover the entire ITU C-band. Nanotechnology was perfected to deliver artificial crystals and take advantage of quantum well QW behaviour. Photodetection was modelled to get optimum SNR and linear response for fast receiver designs.

6   Answers to Exercises

Ex 1

Groups III, IV and V have the photonic materials:
III   Al, Ga, In,
IV   Si, Ge,
V   P, As

Ex 2

Calculate the wavelength associated with a 1 eV energy gap and also for Si and GaAs where it is 1.12 eV and 1.44 eV respectively.

$E_g = hc/\lambda$ gives $\lambda = hc/E_g$

At 1 eV $E_g = 1.6 \times 10^{-19}$ J
Hence $\lambda = 6.6 \times 10^{-34} \times 3 \times 10^8 / 1.6 \times 10^{-19} = 1.24$ μm

At 1.12 eV for Si... $\lambda = 1.24/1.12 = 1.11$ μm

At 1.44 eV for GaAs... $\lambda = 1.24/1.44 = 0.86$ μm

Ex 3

A forward current of 30 mA is injected into a GaAs LED with quantum efficiency 95%. What output light power results? Use data from Ex 2.

GaAs quantum efficiency 0.95

$I = 30$ mA $= 0.03/1.6 \times 10^{-19}$ electrons per second

Optical power $P_o = 0.95(0.03/1.6 \times 10^{-19})1.44 \times 1.6 \times 10^{-19}$
$$= 0.95 \times 0.03 \times 1.44 = 0.04xx \text{ W} = 40 \text{ mW}$$

Ex 4

External quantum efficiency:
$$F = \tfrac{1}{4}(n_2/n_1)^2[1-(n_1-n_2)^2/(n_1+n_2)^2]$$

For GaAs $n_1 = 3.6$ while in air $n_2 = 1$ gives:

$$F = \tfrac{1}{4}(1/3.6)^2[1-(0.36)^2/(2.36)^2] = 1.9\%$$

Ex 5

Find J_t and I_t for a GaAs laser.
L=200x10^{-6} m   w=10x10-6 m   R=0.32 from n=3.6
Γ=0.8 α = 10 cm^{-1} β = 2.0x10^{-4} mA^{-1}

$I_t = (wL/\beta)[\alpha+(1/L)\ln(1/R)]$
$=10^{-5}\text{x}2.0\text{x}10^{-4}/2.0\text{x}10^{-4}[10\text{x}10^2+(1/2.0\text{x}10^{-4})\ln(1/0.32)]$
$=67\text{x}10^{-3} =67$ mA

$J_t = I_t/wL = 67\text{x}10^{-3}/(10^{-5}\text{x}2.0\text{x}10^{-4}) = 3.35\text{x}10^7$ Am^{-2}

For Γ=1: J_t=3.35 kAcm^{-2} and I_t=67 mA

For Γ=0.8: J_t=3.35/0.8=4.19 kAcm^{-2} and I_t=83.7 mA

Ex 6

R increased by 10% due to back-reflection from glass fibre facet. Now R_1=0.32 but R_2=0.352 so $(R_1R_2)^{1/2}$ becomes $(0.32\text{x}0.352)^{1/2}$ = 0.335 rather than 0.32
Now $\ln(1/R)$ = 1.094 rather than 1.139 so I_t falls by 1.041 or to 96% previous threshold.

New I_t = 67x0.96 = 64.3 mA
This will raise output power for same bias current I_b.

Ex 7

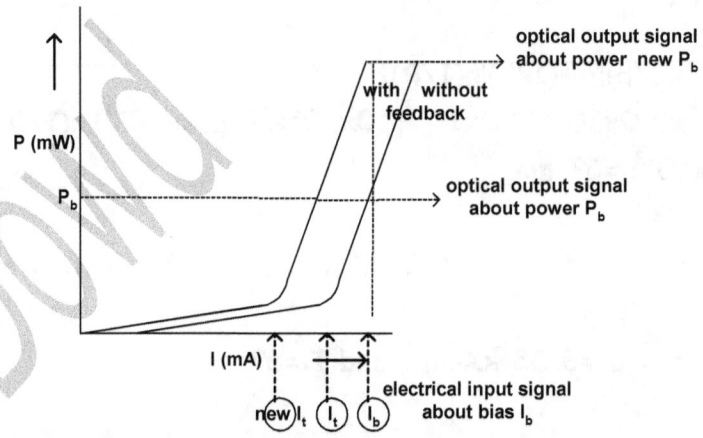

The feedback in Ex 6 has reduced I_t so at the same bias I_b the output signal is about a higher P_b as illustrated.

Note: The reverse occurs with two L-I plots for temperatures T and T' as in Figure 22 so that at the higher temperature the input signal is about a plot with *raised* threshold and "clipping" occurs on the optical output. Draw this case showing input and output modulation.

Ex 8

A typical communications laser emits 10 mW or 10dBm but a high power device is selected for 20 dBm as the link is atmospheric not fibre optic. It lases at 1.5 μm and the semiconductor alloy has average index 3.45 across the lasing spectrum. The device dimensions are stripe w=10 μm, lit depth D=1 μm, chip length L=100 μm. Calculate the photon density s.

$P = s(hc/\lambda)Vc_g$ gives $s = P\lambda/hcVc_g$

$P = 20$ dBm $= 100$ mW $= 0.1$ W
$V = 10 \times 1 \times 100 \times 10^{-18}$ m^3 $= 10^{-15}$ m^3
$c_g = 3 \times 10^8/3.45 = 0.87 \times 10^8$ ms^{-1}

$s = 0.1 \times 1.5 \times 10^{-6}/(6.6 \times 10^{-34} \times 3 \times 10^8 \times 1000 \times 10^{-18} \times 0.87 \times 10^8)$
$= 8.7 \times 10^{33}$ photons/m^3 $= 8.7 \times 10^{15}$ photons/μm^3

Sample Examination Questions

$$c = 3 \times 10^8 \ \text{ms}^{-1}$$
$$e = 1.6 \times 10^{-19} \ \text{C}$$
$$h = 6.6 \times 10^{-34} \ \text{Js}$$
$$k = 1.38 \times 10^{-23} \ \text{JK}^{-1}$$
$$m_e = 9.11 \times 10^{-31} \ \text{kg}$$

A   (a)   In an injection laser the gain and "losses" are related by

$$g\frac{d}{D} = \alpha + \frac{1}{L}\ln\left(\frac{1}{R}\right).$$

Discuss the implications for diode structure taking into account the relationship between g and J_t, the threshold current density.

    (b)   The reflectivity R is increased from 0.36 by 10% due to optical feedback. Calculate the change in I_t, given the following:

$$\alpha = 12 \ \text{cm}^{-1}$$
$$L = 200 \ \mu\text{m}$$
$$I_t = 50 \ \text{mA (without feedback)}.$$

B   For good frequency response in a LED the minority carrier lifetime τ should be small. Discuss the diode *operational* and *design* features by which this is achieved and the limitations on each approach.

In the latter case describe design for various types of confinement.

C   Establish expressions for the signal-to-noise ratio for an APD and a *p-i-n* detector respectively. Consider a transmitted sinusoidal optical waveform.

Show that when large S/N ratio is required of a system it is generally operating near the shot noise limit; and that when small S/N ratio is pertinent the limit is set by circuit noise.

In the APD case show there exists an optimum avalanche gain yielding maximum S/N ratio.

D (a) Show that for a quantum well semiconductor laser the active layer thickness should satisfy

$$L_z < \left(\frac{3h^2}{8m^*kT} \right)^{\frac{1}{2}}$$

in order that only one quantised sub-level should reside within the occupied conduction band. (Symbols have usual meanings).

(b)   What is meant by low-dimensional laser structures?   Describe examples.

(c)

E (a)   (i)   Explain your understanding of the term "internal quantum efficiency" for a LED and how to achieve close to 100% in surface emitters.

(ii)   Describe the double heterostructure in edge emitters and how this impacts on quantum efficiency.

(b)   Consider a surface emitter LED and discuss the factors that influence the recombination lifetime where high-speed response is required.   Show there is a conflict with implications for (a) (i) above.   Show that with higher modulation current the speed improves but at a cost.

F (a)   (i)   Establish the relation for the ratio of spontaneous to stimulated emission coefficients that derives from Planck's and Boltzmann's laws.

(ii) Show that this ratio can be $\sim 10^5$ at optical fibre transmission frequencies but considerably more favourable for coherent stimulated emission at radio frequencies around 100 MHz.

(b)   Describe the effect of temperature on the threshold current for diode lasers and show that a high characteristic temperature T_0 is desirable.   Comment generally on temperature effects in this context.

G (a)   For a photodiode in photoconductive mode, deduce the responsivity-versus-wavelength behaviour in the ideal case.   Carefully plot this result for comparison with a silicon detector explaining the divergences in performance.

(b)   Describe the gain process and related phenomena in APDs for various materials.   Discuss how the noise is quantified.

H   (a)

    (i)   Explain your understanding of the term "internal quantum efficiency" for a LED and how to achieve close to 100% in surface emitters.   25%

    (ii)   Describe the double heterostructure in edge emitters and how this impacts on quantum efficiency.   25%

    (b)

    (i)   Consider a surface emitter LED and discuss the factors that influence the recombination lifetime where high-speed response is required.   25%

    (ii)   Show there is a conflict with implications for (a) (i) above.
Show that with higher modulation current the speed improves but at a cost.   25%

J   (a)

    (i)   Establish the relation for the ratio of spontaneous to stimulated emission coefficients that derives from Planck's and Boltzmann's laws.   30%

    (ii)   Show that this ratio can be $\sim 10^5$ at optical fibre transmission frequencies but considerably more favourable for coherent stimulated emission at radio frequencies around 100 MHz.   30%

    (b) Describe the effect of temperature on the threshold current for diode lasers and show that a high characteristic temperature T_0 is desirable. Comment generally on temperature effects in this context.   40%

K   (a) For a photodiode in photoconductive mode deduce the responsivity-versus-wavelength behaviour in the ideal case. Carefully plot this result for comparison with a silicon detector explaining the divergences in performance.   60%

    (b) Describe the gain process and related phenomena in APDs for various materials. Discuss how the noise is quantified.   40%

Appendix 1   Periodic Table of Elements

Group:

I	II											III	IV	V	VI	VII	VIII
1																	18
1 H 1.008	2											13	14	15	16	17	2 He 4.0026
3 Li 6.94	4 Be 9.0122											5 B 10.81	6 C 12.011	7 N 14.007	8 O 15.999	9 F 18.998	10 Ne 20.180
11 Na 22.990	12 Mg 24.305	3	4	5	6	7	8	9	10	11	12	13 Al 26.982	14 Si 28.085	15 P 30.974	16 S 32.06	17 Cl 35.45	18 Ar 39.948
19 K 39.098	20 Ca 40.078	21 Sc 44.956	22 Ti 47.867	23 V 50.942	24 Cr 51.996	25 Mn 54.938	26 Fe 55.845	27 Co 58.933	28 Ni 58.693	29 Cu 63.546	30 Zn 65.38	31 Ga 69.723	32 Ge 72.63	33 As 74.922	34 Se 78.96	35 Br 79.904	36 Kr 83.798
37 Rb 85.468	38 Sr 87.62	39 Y 88.906	40 Zr 91.224	41 Nb 92.906	42 Mo 95.96	43 Tc (98)	44 Ru 101.07	45 Rh 102.91	46 Pd 106.42	47 Ag 107.87	48 Cd 112.41	49 In 114.82	50 Sn 118.71	51 Sb 121.76	52 Te 127.60	53 I 126.90	54 Xe 131.29
55 Cs 132.91	56 Ba 137.33	57-71 *	72 Hf 178.49	73 Ta 180.95	74 W 183.84	75 Re 186.21	76 Os 190.23	77 Ir 192.22	78 Pt 195.09	79 Au 196.97	80 Hg 200.59	81 Tl 204.38	82 Pb 207.2	83 Bi 208.98	84 Po (209)	85 At (210)	86 Rn (222)
87 Fr (223)	88 Ra (226)	89-103 #	104 Rf (265)	105 Db (268)	106 Sg (271)	107 Bh (270)	108 Hs (277)	109 Mt (276)	110 Ds (281)	111 Rg (280)	112 Cn (285)	113 Uut (284)	114 Uuq (289)	115 Uup (288)	116 Uuh (293)	117 Uus (294)	118 Uuo (294)

* Lanthanide series	57 La 138.91	58 Ce 140.12	59 Pr 140.91	60 Nd 144.24	61 Pm (145)	62 Sm 150.36	63 Eu 151.96	64 Gd 157.25	65 Tb 158.93	66 Dy 162.50	67 Ho 164.93	68 Er 167.26	69 Tm 168.93	70 Yb 173.05	71 Lu 174.97
# Actinide series	89 Ac (227)	90 Th 232.04	91 Pa 231.04	92 U 238.03	93 Np (237)	94 Pu (244)	95 Am (243)	96 Cm (247)	97 Bk (247)	98 Cf (251)	99 Es (252)	100 Fm (257)	101 Md (258)	102 No (259)	103 Lr (262)

Groups III, IV and V have the photonic materials:

III   Al, Ga, In,

IV   Si, Ge,

V   P, As

Appendix 2   Table of Fundamental Constants

Speed light in free space   c 3×10^8 ms^{-1}

Plank's constant   h 6.6×10^{-34} Js

Electronic charge   e 1.6×10^{-19} C

Electron volt   1 eV = 1.602×10-19 J

Mass electron   m$_e$ 9.1×10^{-31} kg

Boltzmann constant   k 1.38×10^{-23} JK^{-1}

Permittivity free space ε_0 8.85×10^{-12} Fm-1

Permeability free space μ_0 1.26×10^{-6} Hm-1

Solar constant surface earth   495 Wm-2

Solar λ maximum intensity   500 nm